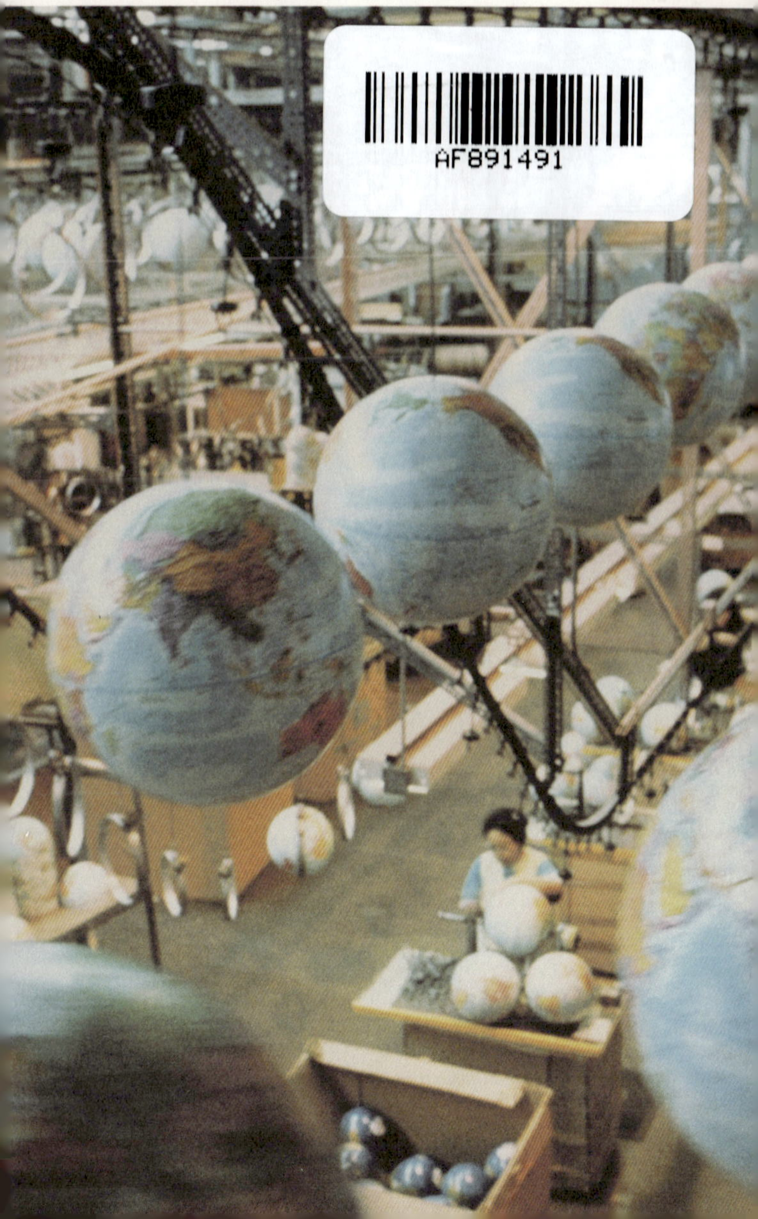

Overleaf: Andy Sacks, *Globes High Up on Transport System Above Factory Floor*, 1995. (photograph Andy Sacks, courtesy Getty Images)

Terrestrial Tales
100 + Takes on Earth

Marc Angélil and Cary Siress
with
Marcin Ganczarski, Tobias Klauser,
Ciro Miguel, Ellen Reinhard, Jenny Rodenhouse, David Roth,
Julian Schubert, Elena Schütz, Leonard Streich

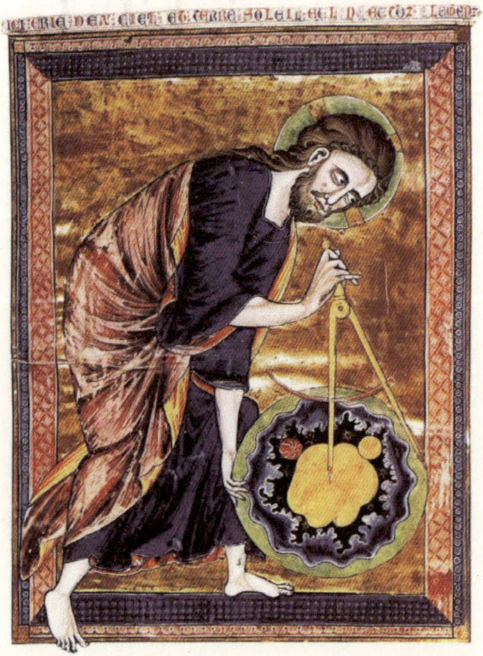

Frontispiece of the *Codex Vindobonensis 2554* from the early 13th century showing God as geometer or craftsman of the world (courtesy Österreichische Nationalbibliothek, Vienna)

Terrestrial Tales

Who does the Earth think it is?

Gilles Deleuze and Félix Guattari, 1980[1]

Where on Earth do you begin a story about the Earth? For many, the story of divine creation would seem to have closed the book on the question of beginnings once and for all. For others, the story of globalization would seem to have set us on a path from which there is no return, marking the onset of an increasingly human-made world. Each story in its own way bears on the making of our place in the larger order of things – Earth determined by a higher power and Earth determined by humankind.

As a matter of fact, this is how it all started, at least for our collaboration at ETH Zurich. In an early lecture of the introductory course for architecture design, the opening slide showed an image of God as supreme craftsman

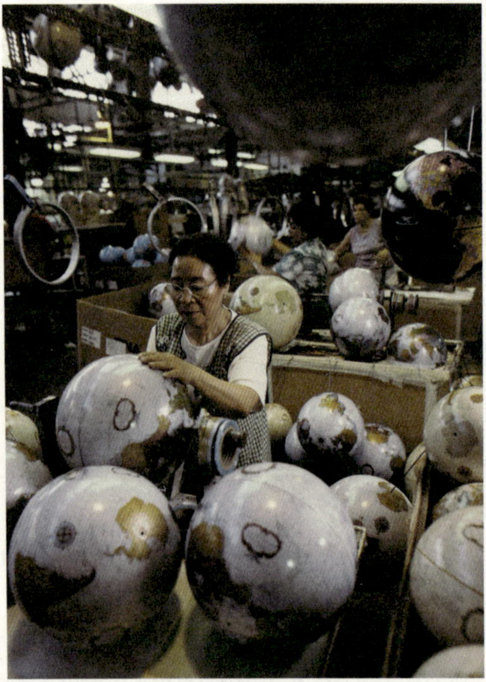

Migrant worker assembling
a globe in a factory in
Broadview, Illinois, 2001
(photograph Andy Sacks,
courtesy Getty Images)

with compass in hand administering the final touches on His creation.[2] What better place to start a foundational story in a *Grundkurs* (Architecture 101) than with the ur-scene itself showing the primal act of making at the very moment when unformed matter is transformed into the consummate form of the Earth. In a later lecture, an image was shown of a migrant worker in a factory putting the finishing touches on one of the many globes produced during her shift on the production line.[3] What better place to begin a story on world development than there where it is produced day in and day out by an anonymous workforce paid close to nothing for its labor. The juxtaposition of these two images, at first perhaps nothing more for students than a chance encounter, was meant to strike a dialogue between takes on world-making that are poles apart – one take until now reserved solely for theological interpretation, the other implying the extent to which the Earth in its totality has become ready-to-hand as a world commodity.

In time, more and more images would be collected for research on world-making as an ongoing project of human design, forming a cabinet of curiosities of sorts that made us all too aware of what is at stake in that very project. This collection is peculiar, indeed, inasmuch as it sets marvels next to monstrosities, the mundane next to the exotic, and the scientific next to the artistic – all laden with their own agendas. The material gathered opened up a discourse on stories that concern not only how we see the world, but also how our world is constructed accordingly as a compendium of competing 'terrestrial tales' or operative narratives that maintain our worldview (*Weltbild*), our concept of humanity (*Menschenbild*), as well as our own self-image as distinct subjects (*Selbstbild*). As if having entered a hall of mirrors, the efforts made to probe such constitutive narratives inevitably brought us face to face with countless takes on planet Earth that – as so many reflections on where we collectively stand – have made the world in its conception as well as in its concrete manifestation what it is today, for better or worse.

To expand the collection and gain an even broader overview of contemporary perspectives on world-shaping agencies and their increasing entanglement, a "Call for

Globes" went out to faculty and students of ETH Zurich in early 2019. The responses were wide-ranging, coming as they did from disparate disciplines, whether from the sciences addressing the subject matter of climate change, from planning and architecture raising questions about global urbanization, or from the arts and humanities reflecting on processes of planetary transformation. Soon thereafter, other universities around the globe answered the call for globes, bringing ever more stories into the mix.

In sum, all the entries formed the truly unanticipated collection of the exhibition *Terrestrial Tales: 100+ Takes on Earth* assembled in the main hall of ETH Zurich in October 2019. Visitors can look down from the galleries above on an array of screens arranged on a dark blue carpet. While perhaps exhilarating for some, this particular arrangement implicates all viewers in what has become our commonly objectified view of the world from afar, even though all of us know deep down that we are manifestly earthbound, obligated to cooperation, and co-responsible for the entirety of the environments we make and inhabit.

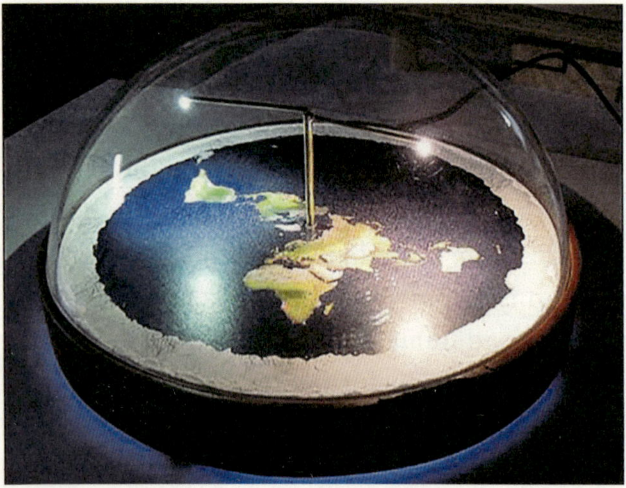

Flat earth model by Chris Pontius,
for purchase on his personal website
www.flatearthmodels.com, 2017

Spins of the Globe

Stuffed with diverse curiosities as it is, the *Wunderkammer* of assorted takes on Earth includes items that could just as well be from another world. Take, for instance, the maps and models made by members of the Flat Earth Society used to articulate their contentious conception that the world is a disk, which, as a critique of the dominant globe paradigm, revives an ancient cosmological model to give it a new lease of life in our time. The society uses an array of promotional techniques for disseminating its worldview, ranging from conferences and publications to an official website and videos. One YouTube video in particular shows a scaled mock-up of Earth as a circular plane beneath a rotating celestial dome made of plexiglass. For added effect, the whole device is equipped with blinking lights to highlight cities as well as the stars above—an item, by the way, that can be purchased from an online shop.[4]

Another example of a controversial take on Earth comes by way of a Chinese map, which, although dating from 1763, is allegedly a reproduction of an older map from 1418

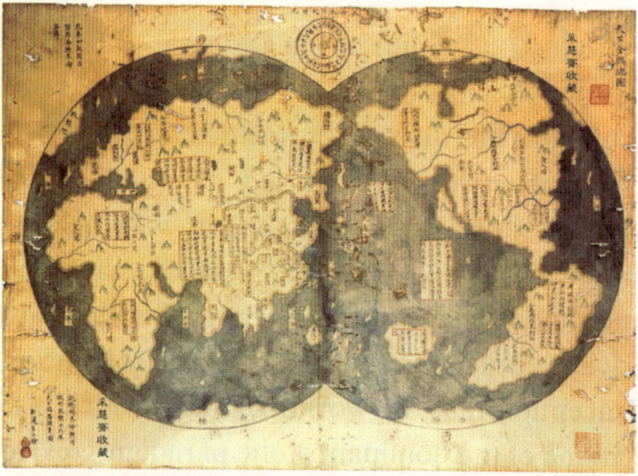

Controversial Chinese map from 1763
reported to be a reproduction of a
1418 map showing the outline of the
Americas, unveiled in 2006

that renders the Americas quite accurately, and this some 70 years before their renowned discovery by Christopher Columbus. The map's authenticity is still refuted by skeptics refusing to believe that the Chinese were the first to put the New World on the map. This skepticism is perhaps understandable because, if genuine, the map would "topple one of the central tenets of Western civilization: that Europeans were the first to sail around the world and discover America."[5] This would certainly explain the storm that erupted when the map was made public in 2006 by a prominent Chinese attorney and collector. The issue, however, was not necessarily its credibility per se, but rather its potential to replace the longstanding story of a Eurocentric world with one recentering the Middle Kingdom in the scheme of things.

In a fundamentally other, but no less intriguing, way is that particular take on Earth by Buckminster Fuller and John McHale showing a vertical section reaching from the depths of the oceans all the way to the Sun to demarcate the extended reach of human intervention. Entitled "Man's Increasing Vertical Mobility," the drawing was presented in a lengthy proposal to

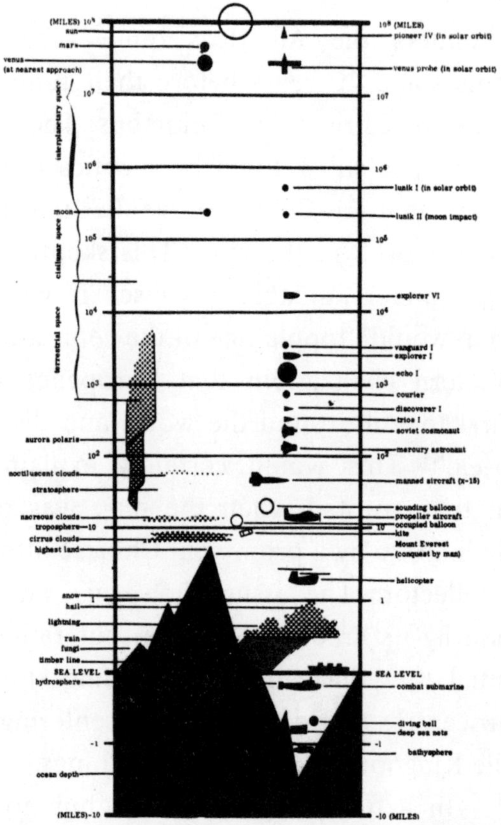

Buckminster Fuller and John McHale,
sectional drawing cut through the full height
of the zone of human intervention, 1963
(courtesy Buckminster Fuller Institute)

the International Union of Architects (IUA) in the early 1960s, the intention being to overhaul architecture education worldwide by bringing it up to date relative to the broad socio-technical transformations underway. This and other like diagrams were meant to reveal agents of change not only operating in the visible realm of everyday life, but also those in more remote regions—whether below sea level or in outer space—that are nonetheless active as drivers of modernization. Of concern was the design of future-oriented strategies for boosting the efficiency of what Fuller called "Spaceship Earth," strategies that in the best of cases would also lead to the more equitable distribution of resources. In place of 'superficial functionalism' and 'aloof formalism' that were the mainstay of traditional architecture education at the time, Fuller and McHale proposed their so-called "World Game," in which students and faculty from diverse disciplines would playfully explore design scenarios that work at the planetary scale to the benefit of the planet.[6]

Whereas such examples register the often conjectural, occasionally scandalous, and sometimes radically projective nature of stories

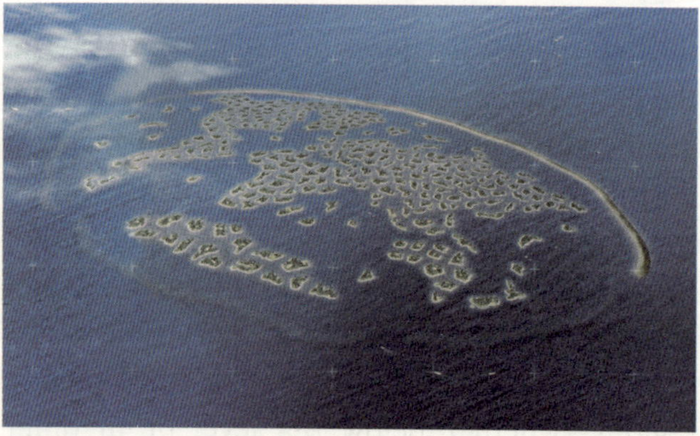

Rendering of *The World* just off the shore of Dubai by Nakheel Properties, at the time one of the most spectacular portfolio holdings of Dubai World, ca. 2010 (courtesy Nakheel Properties)

about Earth, there are others that combine all of these traits in the effort to make their claims hold. *The World* in Dubai is one such case. As if taken straight from a fairytale, the project is conceived as an artificial archipelago of three hundred islands built just offshore that together form the familiar contours of the world as it appears on a map. In what is portrayed as an innocuous image drawn on the surface of the sea, one that incidentally is meant to be seen via Google Earth, is more than a mere gimmick. In fact, *The World*, a real estate venture unveiled in 2003, is integral to Dubai's carefully crafted self-marketing campaign intended to give the emirate geopolitical weight. The promotional claim by Nakheel Properties that "Dubai puts the World on the Map" just as well puts Dubai on the map, or better, makes it a patent icon on the neoliberal globe, giving rise to a city validated by the world-image itself. Though most of the islands remain undeveloped, the flat earth motif has served its purpose well in turning the emirate-city into a prized commodity on the world market.

There are still other takes that are diametrically opposite to those born by flights of fancy. The work of conceptual artist Karin

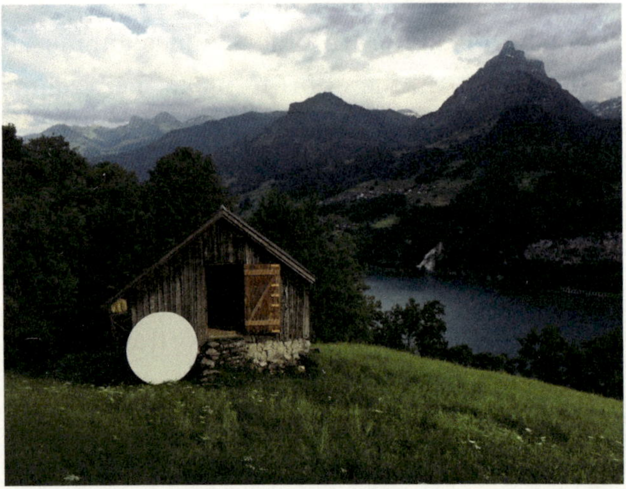

Gebrauchs-Bild by Karin Sander,
Atelier Amden, 2014-2019
(courtesy Karin Sander)

Sander could not be further from the world of fabricated images, insofar as her *Gebrauchs-Bilder* (use- or patina-paintings), for instance, are produced by happenstance; they are the result of circumstantial processes on site that leave their traces on a canvas left in a particular location. Because such indexical installations bear on transformations over time, they are always in progress, "open works" in the truest sense.[7] For one such use-painting, initiated in 2014 for the Atelier Amden, a circular canvas was placed in a bucolic setting on the outskirts of the metropolitan region of Zurich.[8] This installation could be read as a flattened globe on which the surrounding environment works in a most literal way. In place of offering one more take on Earth, Sander seems more concerned with registering how the Earth takes on the canvas, layer upon layer and without end, as a bundle of agencies in its own right.

And the stories go on. In the attempt to fathom what our planet has indeed become today and to tell its story, the term "geostory" has figured as a key concept in contemporary discourses on the state of the world environment. Bruno Latour, in particular, makes reference

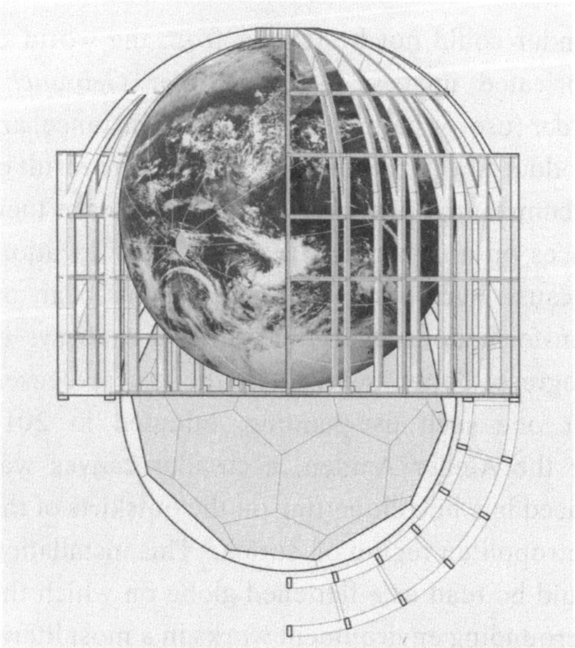

Drawing by Design Earth (Rania Ghosn and El Hadi Jazairy) for a 2017 installation entitled "Blue Marble Circus" inscribing a blue shrink-wrapped globe within the spherical volume of the Pantheon to represent the plastic footprint of humanity, an impact that remains outside our geographical imagination (published in *Geostories: Another Architecture for the Environment*, 2018)

in his 2013 Gifford Lectures to what he calls our "common geostory" to include not only those stories that humans tell about Earth, but also the unfolding narrative of Earth as it plays out in ever more volatile ways due to our collective impact on its behavior.[9] Latour recasts the terrestrial domain, in other words, as the prime agent in shaping the history of humanity and its relation to the planet, thus making the sum of human actions – those in the past and those to come – unavoidably "geostorical" by nature. Acknowledging our impact on planetary ecologies, he asserts that we are urgently in need of new stories to make the Earth public, that is, to make the state of our planet a matter of concern for all.

Building on this very logic, the architecture collective Design Earth (Rania Ghosn and El Hadi Jazairy) has crafted a series of novel terrestrial portraits bearing on our roles in world-making.[10] In *Geostories: Another Architecture for the Environment*, the architects deploy design as an agent to represent, in a political sense, "those things, spaces, and scales that are erased from the geographic imagination" to make visible the variegated ways of how we

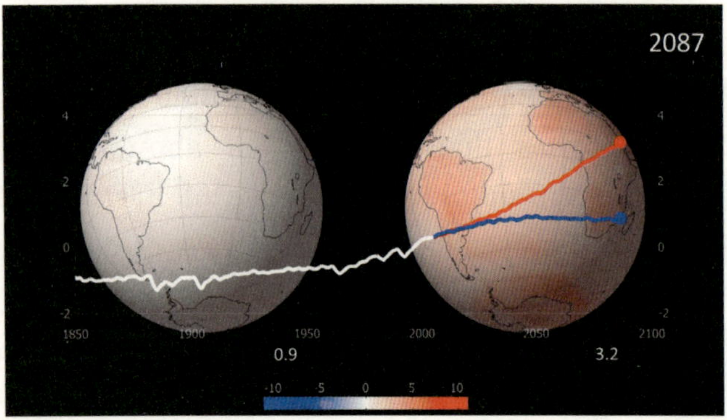

Visualization of two climate scenarios that project a +1°C-world and +4°C-world by 2100; prepared by the Institute for Atmospheric and Climate Science, ETH Zurich, 2012 (courtesy Reto Knutti, Oliver Stebler, Jan Sedlácek, Urs Beyerle)

have altered the Earth and, through design propositions, to speculate on more accountable ways of inhabiting "a damaged planet."[11]

From positions such as these, it would seem that Earth itself is due a more representative voice in its own story. And to varying degrees, this is the very ambition of the natural sciences, which recently have found themselves deeply entangled in the unfolding drama of Earth's story, as discipline after discipline attempts to disclose anthropogenic alterations of the environment at all scales. Whether addressing atmospheric greenhouse accumulation and its effect on global warming, the split of the polar vortex into separate air masses that affects surface weather, or the increasing frequency of tropical cyclones in the Asia-Pacific region and their growing impact on coastal zones, such scientific studies cannot help but become political, especially in an age when objective fact is viewed by some with skepticism as one of many possible takes. Nonetheless, such research and the blaringly obvious that comes through all those sets of figures, charts, animations, scenarios, and the like, have served to gather disparate interest groups around the common cause of drafting

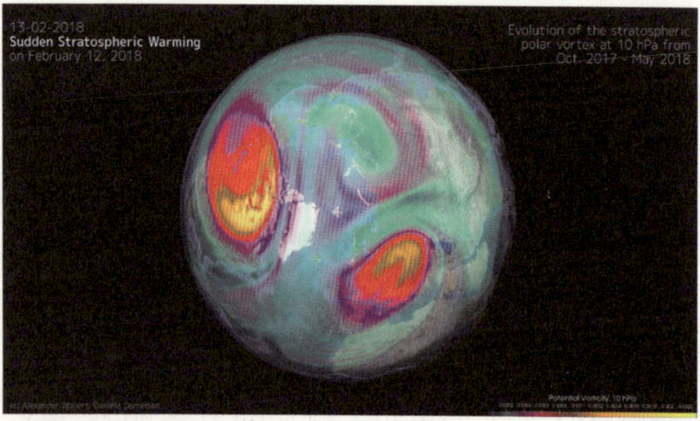

Digital model showing a split of the polar vortex that in 2018 displaced cold air from the North Pole to lower regions of the planet, a rare atmospheric event known as Sudden Stratospheric Warming (SSW); prepared by the Institute of Atmospheric and Climate Science, ETH Zurich, 2018 (courtesy Daniela Domeisen and Alexander Wollert)

binding policy stipulations to mitigate further planetary damage.

The World as Mirror

Just as the stories that make up this compendium are meant to mirror the world, so too does the world mirror our cumulative impact on it. Although the globe might seem to imply a unified space, our representations of the world, the way we act upon the world, and the way the world responds to our actions have in reality become mutually disjunctive, out of sync and increasingly at odds with any attempt to patch things back together. This being the case, there have been many others nevertheless who have likewise called attention to just this predicament.

It is worth recalling Martin Heidegger's 1938 essay "The Age of the World Picture," in which he meditated on the world's diminished status as mere image, as something set off apart from us to be mastered.[12] The core of his argument is that the world has been too frequently looked at for its instrumental value alone, appraised, that is, above all for what it

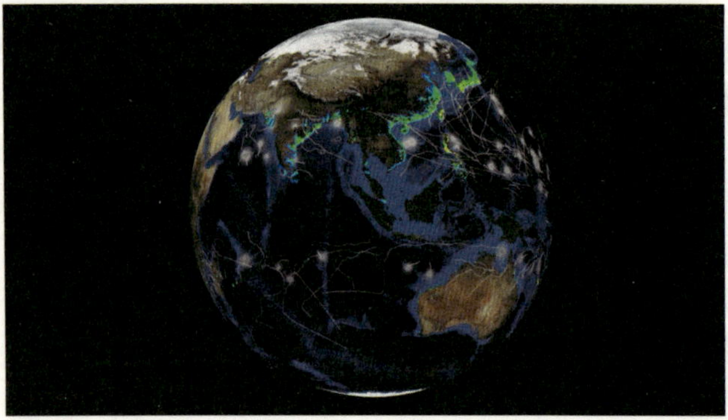

Still from a digital animation showing tropical cyclones in the Asia-Pacific region from 1987 to 2016 and their cumulative impact on coastal regions; prepared by the Institute for Environmental Decisions, ETH Zurich, 2016 (courtesy David N. Bresch and Tarun Chadha)

offers up as a 'standing reserve' to be exploited for political gain. When the world is taken as 'picture', representation is conquest. Many of the world-images that have proliferated to this end reveal just how ingrained this mindset really is and to what extent exploitation can and will be pursued. Although Heidegger appealed for a more profound worldview that might discredit superficial caricatures of the world and, by extension, the actions they legitimize, the now pervasive tendency to level everything into authoritative and alluring images has literally flattened the world in the name of progress.[13]

Building on Heidegger's work, Peter Sloterdijk, in the second volume of his *Spheres* trilogy entitled *Globes*, goes even further by claiming that the globe became the powerful representation that it still is today primarily because it signified, for those who invented it, that any location on Earth is equally accessible and exploitable.[14] This lesson of the globe made the planet legible as the "pictorial and conceptual embodiment of the whole," an ideally unified space in its spherical purity, but nonetheless seen as readily available for the taking.[15] It is this representation of the planet's geography that, according to Sloterdijk, signaled the advent of what Heidegger called the "age of the world picture."[16] Whatever could be projected onto the surface of the globe with a degree of accuracy could be considered sufficiently true. Admitting, however, that the world we have created is far from ideal, mankind has settled for ersatz worlds crafted at the expense of the real one. The container way of life perpetuated by our means and ends has over time drawn everything inwards to what Sloterdijk labels "the world interior of capital."[17] Once Earth had been transformed into the globe, it provided the grounds for conquest and speculation, which would give rise to the integrated world economy. From then on, "social life could only take place in an expanded interior, a domestically and artificially climatized inner space"—the ideal hothouse built both by and for capital.[18]

In response to Sloterdijk's deliberations on our world predicament, Bruno Latour, again, in his own way has reframed debates on the causes and effects of terrestrial crises by calling attention to the lack of an orienting worldview or, for that matter, the lack of a collectivizing concern regarding our rather tenuous condition on Earth. Latour and his team at the Paris Institute of Political Studies (Sciences Po) have used a range of media to re-present the world more inclusively in terms of human and nonhuman interrelations that make up the zone of the planet that *all* inhabit. In the 2016-17 performance lectures entitled *Inside*, Latour challenged the common notion of society and nature as discrete realms. Instead, as the name of the presentation made clear, he insisted that both must be seen as internal to the other.[19] By this he meant that human and nonhuman actors have always been immersed in what he called the "critical zone," that is, never outside the thin inhabitable layer of the Earth. Though we still imagine the world as that divine *Blue Marble* seen from afar, down here it has become a thoroughly "Anthropogenic Earth" – the name given by Alexandra Arènes to one of her visualizations made for *Inside*. Her take on Earth presents it as one would document the scene of a crime. The image shows a rather lumpy amalgam of urbanized patches, excavated topographies, infrastructure networks, wasteland zones, altered oceanic currents, toxic clouds, and the like. Considering that every such formation, as ominous as it is, actually manifests the fallout

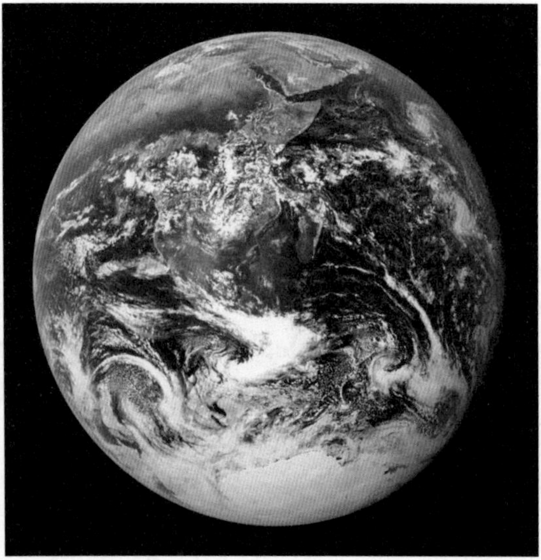

The Blue Marble, photograph of Earth taken by the Apollo 17 crew on December 7, 1972 (courtesy National Aeronautics and Space Administration)

of particular decisions made, particular interests served, particular powers in play, particular techniques deployed, and particular desires pursued, it is as if we have chosen to remain oblivious to the effects of our actions. In having taken flight from Earth in this way, the question we must eventually answer, as Latour reminds us, is "Where do we land?"[20]

Environments in the Making

The aggregate worlds that we have constructed and our ways of inhabiting the planet have dire environmental consequences globally. This seems to sum up rather neatly our current dilemma, save for the fact that it is premised on the reductive standoff between society and nature. Here, nature (as container, as resource, as sublime other) is externalized as a conditioning frame of life increasingly subjected to the whims of human activity. This view has given rise to the prominent narrative of the "Anthropocene," a new era of human-induced transformations of Earth's ecosystems.[21] According to this take, our modes of environment-making have reached

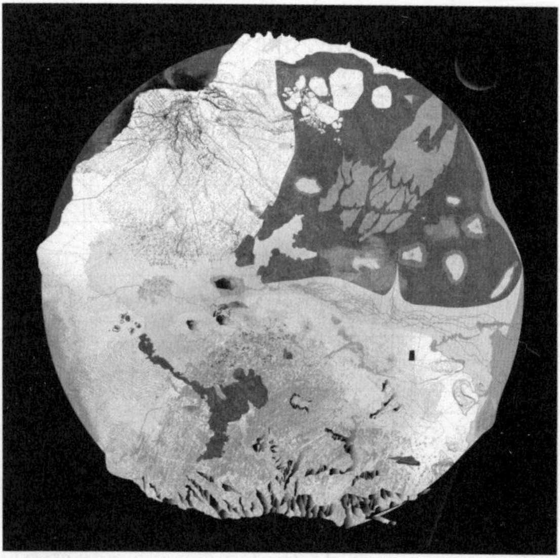

Anthropogenic Earth, visualization by Alexandra Arènes from 2015 showing an interpretation of human-induced transformations of the planet, produced for Bruno Latour's lecture performances *Inside*, 2016-17 (courtesy Alexandra Arènes, Atelier SOC)

such a degree of impact that we now constitute a full-fledged geological force capable of altering the sum of planetary processes, with climate change generally cited as the most arresting effect of society's domination of nature. As it stands from this perspective, "human enterprise is set against the great forces of nature."[22]

Considering, however, that urbanized and natural environments are becoming ever more entwined in mutually destructive relationships, it might be more accurate to speak of what geographer Erik Swyngedouw has identified as the "Urbicene" to implicate urban development as the primary site of global anthropogenic change.[23] According to this take, which acknowledges the thorough urbanization of world society, environment-making must be understood at the global geographic scale to include not only cities and their agglomerations, but also vast productive landscapes, mountain ranges, stretches of forests, and open ocean territories, not to mention the logistics networks running through all of them. Given the extent that urbanizing processes "have become a basic parameter for planetary social and environmental relations," and considering that there is no longer an 'outside' to these processes, nothing has been left untouched by the advance of man-made environments, which continue to reorder human and nonhuman relations in drastically uneven ways.[24] As it stands from this perspective, urbanization accelerates the entanglement of socio-ecological phenomena at all scales.

Still, taking into consideration that urbanization is the prime expression of advanced capitalist accumulation, other scholars such as environmental historian Jason W. Moore have addressed our common predicament with the more connective concept of the "Capitalocene."[25] With respect to this particular take, one critical of the long-standing separation of humankind from the rest of nature, environment-making must be viewed as a cascade of "processes and relations, through which particular bundles of human and extra-human nature flow, upon which these bundles act, and re-form as they act."[26] Environment-making, in short, can no longer be framed by the strictly opposing terms of 'man' versus 'nature'. Rather, it must be reconsidered in view of the privileged roles given to capitalism in the coevolution of society, technology, and nature. Just as well, environment-making must be reconsidered as the prime channel for capital in leveling everything to the bottom line of a cheap resource to be exploited.[27] As it stands from this perspective, capitalism does not operate outside of and on nature but *through* nature to steer the development of socio-technical-ecological systems in accordance with vested interests, implying that the production of space encompasses "the production of nature."[28]

These takes on Earth (Anthropocene, Urbicene, Capitalocene) converge on the diagnosis that the world ecology, planetary urbanization, and the globally integrated market are locked in an intimate symbiosis that is far from stable. This diagnosis in mind, environment-making in all its variegated forms has generated such degrees of fragmentation as to fundamentally confound any coherent representation of our shared condition. So, a provocative question like "Who does the Earth think it is?" could just as well be addressed to us: "Who do we think we are?" If left to our current devices, then a more inclusive and sustainable response would seem to be foreclosed from the outset. Yet, if we were to ever truly face up to the incontrovertible fact that Earth is the only

Drawing by architect Angelo Bucci entitled *The Thin Layer* depicting Earth's boundary as a shallow zone of approximately 3.5 kilometers that constitutes the inhabitable realm of the planet; what appears as a single line is actually comprised of two lines, the inner and outer circles of the thin layer (courtesy Angelo Bucci)

habitat available, then we might finally consider it as a de facto commons and accordingly treat terrestrial environment-making as a pressing matter of collective concern. This said, to take on Earth in more responsive ways is to take on ourselves as stakeholders responsible for creating new ways of thinking and working through the 'inside' by putting the production of space and the production of nature on par as equally political agendas. Admittedly, this is an imperfect hope at best and may strike some as just another take. But 'here', we are. And it is due time that we all get down to Earth.

Marc Angélil and Cary Siress

100 +
Takes on Earth

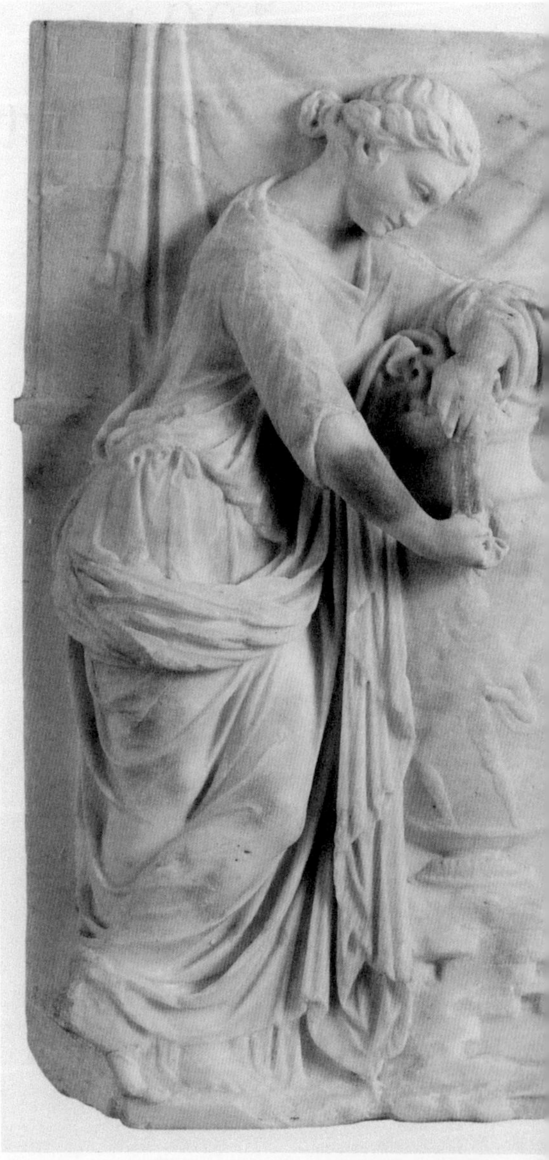

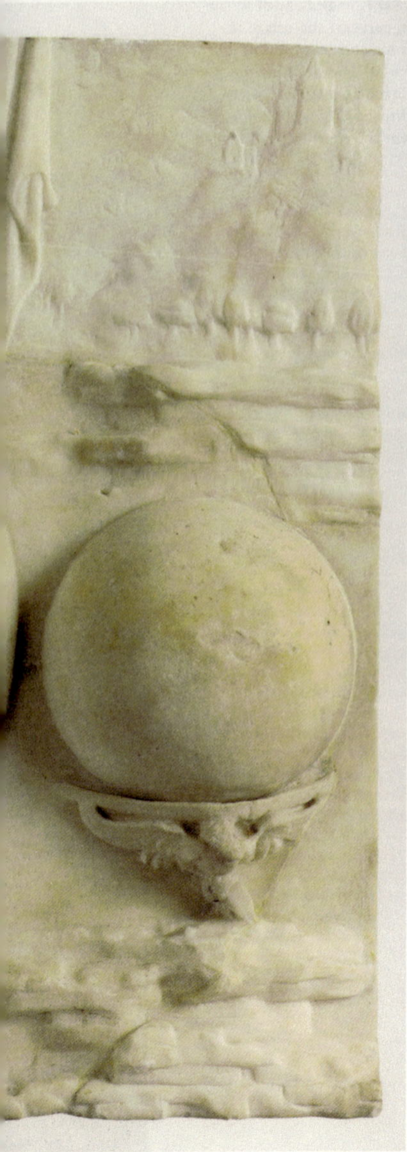

Italian marble relief depicting Urania, the Muse of Astronomy in Greek mythology, gesturing toward the celestial orb as if to foretell the future of the universe by the arrangement of the stars, a narration from antiquity that has been commonly interpreted as one concerning the creation of the cosmos, ca. 1st century AD; the original staff in her hand that was pointing toward the globe is missing. (courtesy Musée du Louvre, photograph ©Musée du Louvre, Dist. RMN-Grand Palais / Anne Chauvet)

The *Farnese Atlas*, a 2nd century AD Roman copy of a Greek sculpture, depicts the Titan carrying the orb of the heavens on his shoulders, with iconic constellations of antiquity placed in relation to circular lines defining the celestial equator, the tropics, the Arctic Circle, and so forth, manifesting a work at the threshold of a mythological and scientific conception of the world. (courtesy National Archaeological Museum Naples / © Alamy)

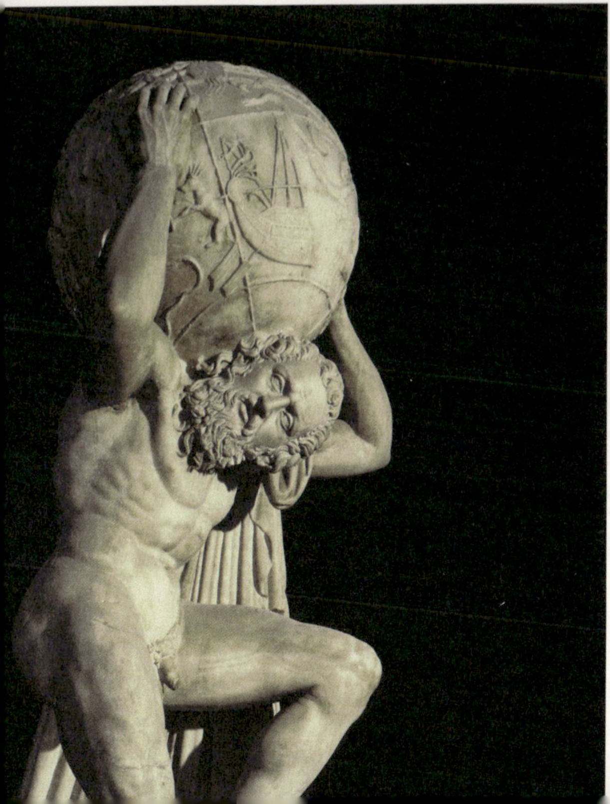

Sectional drawing of the Pantheon (ca. 113-126 AD) in Rome by Francesco Piranesi from 1790; the interior volume, modelled around a sphere, was possibly meant to represent the arched vault of the heavens punctured by the oculus symbolizing the Sun.[1] (engraving from Francesco Piranesi's *Raccolta de' tempj antichi,* vol. I, 1780)

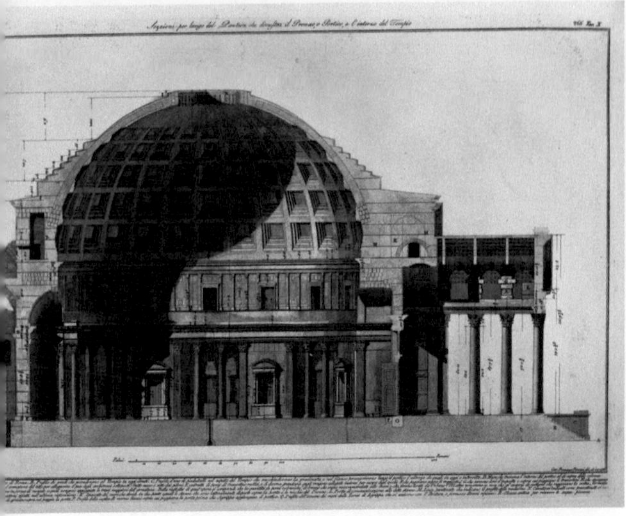

16th century manuscript illustration of Bartholomaeus Anglicus's description of the terrestrial waters in *On the Properties of Things* from 1240; using the then-familiar 'T and O' type of map, the disc of the world is rendered with oceans circumscribing the known world and connected by the main water bodies separating Asia (above), Africa (lower right), and Europe (lower left). (courtesy Bibliothèque Nationale de France and © akg images)

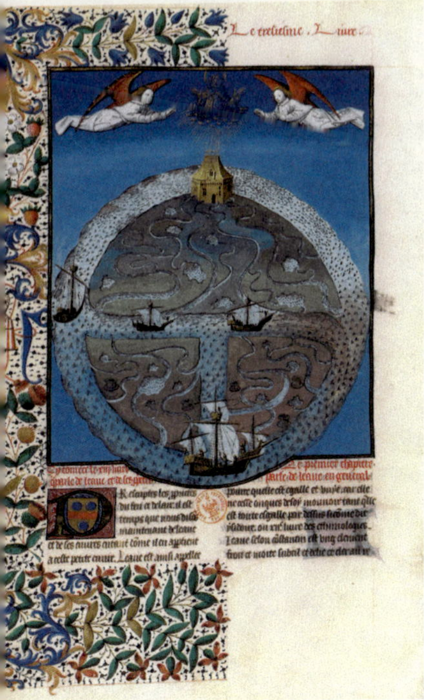

The Behaim-Globe, or "Erdapfel" (literally "Earth Apple"), created by geographer Martin Behaim from 1490-1493 and the oldest known terrestrial globe to date, is a composite of fact and conjecture in that it was fabricated at a time when there was no conclusive scientific proof that the Earth was indeed a sphere; as for those blank spaces on the surface of the globe, they still had to be imagined.[2] (courtesy Germanisches Nationalmuseum Nürnberg and © akg images)

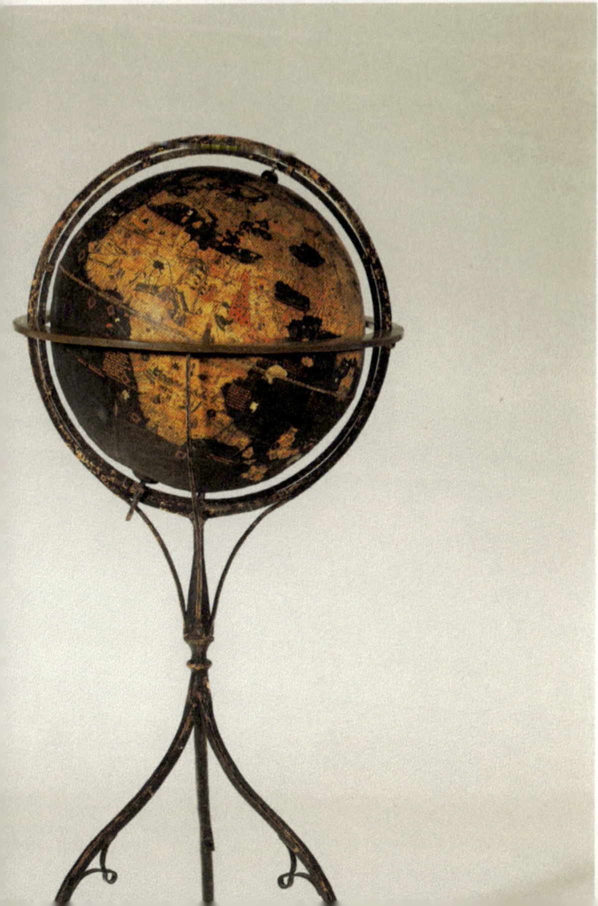

External panels of the triptych *The Garden of Earthly Delights* by Hieronymus Bosch from ca. 1505 representing the third day of the world's creation; whereas the mood of the outside is rather somber, with the scene muted in its colors, the interior setting of the world is vibrant as the narrative progresses from the original sin, to the subsequent proliferation of earthly pleasures, to the eternal torments of hell.[3] (courtesy Museo del Prado and © akg images)

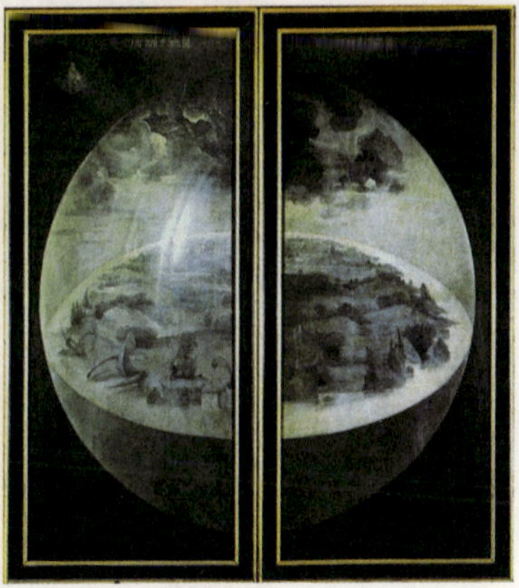

Tapestry by Bernard van Orley from ca. 1530 entitled *Earth Under Protection of Jupiter and Juno* – an allegorical portrait of the Portuguese King John and his wife Catherine as gods reigning over their empire early on in the Iberian colonization of the African continent. (courtesy Getty Images)

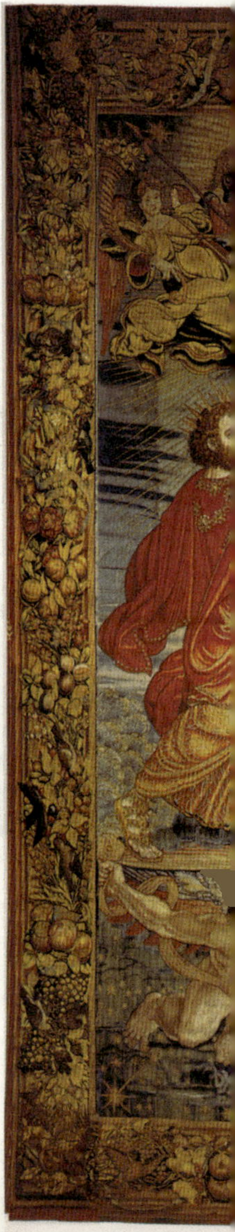

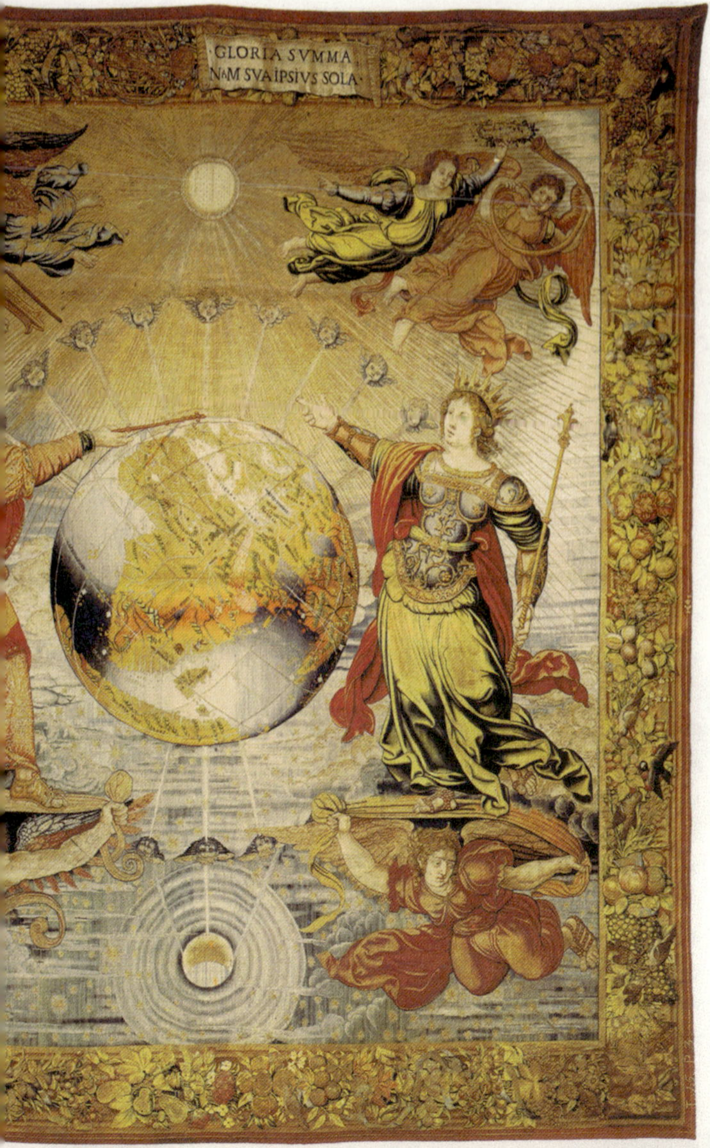
GLORIA SVMMA
NAM SVA IPSIVS SOLA

Reportedly stolen in the course of the 1712 religious war, the St. Galler Erd- und Himmelsglobus from 1571-84 showing the celestial and terrestrial spheres collapsed onto the surface of the globe is currently part of the permanent collection of the National Museum in Zurich; an exact replica of the globe was unveiled in 2009 and given by Zurich to St. Gallen amidst legal action taken by the latter to have the original returned to its rightful place.[4] (photograph Christian Beutler/ Keystone, courtesy Landesmuseum Zürich)

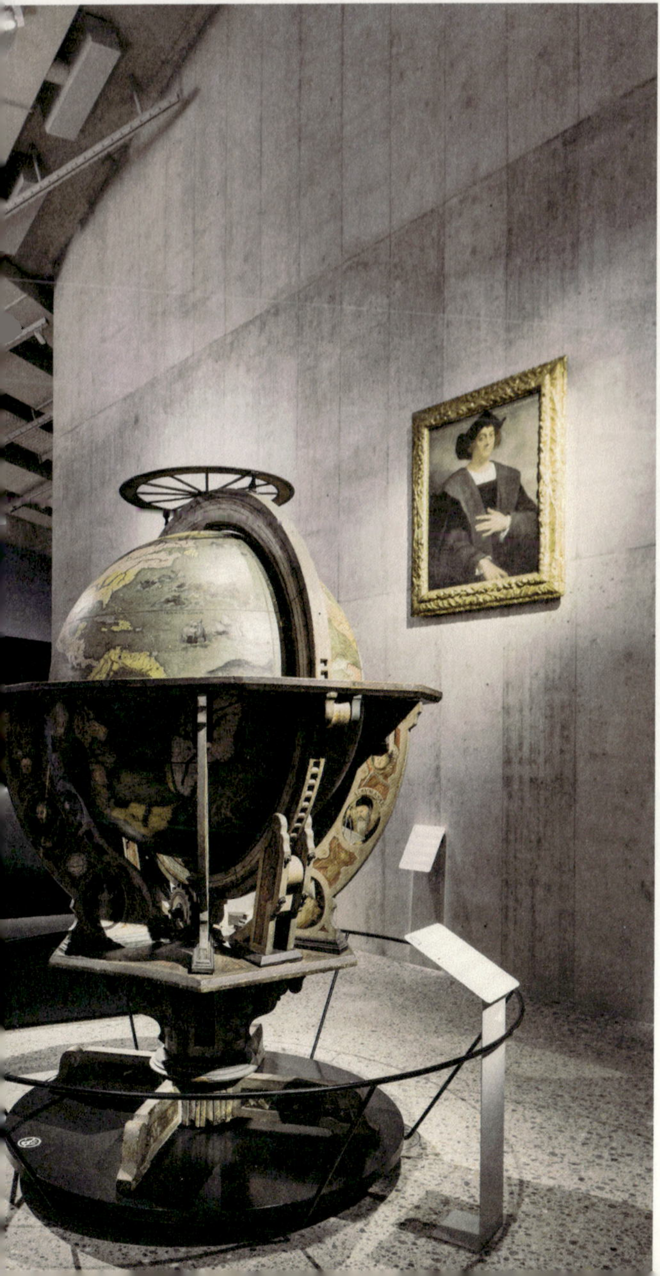

Detail from the allegorical panel painting of Queen Elizabeth I as seen in the Woburn Abbey version of the *Armada Portrait*, her hand on a diminutive globe to symbolize England's dominion of the world; painting attributed to George Gower, 1588.[5] (courtesy Woburn Abbey and © akg images)

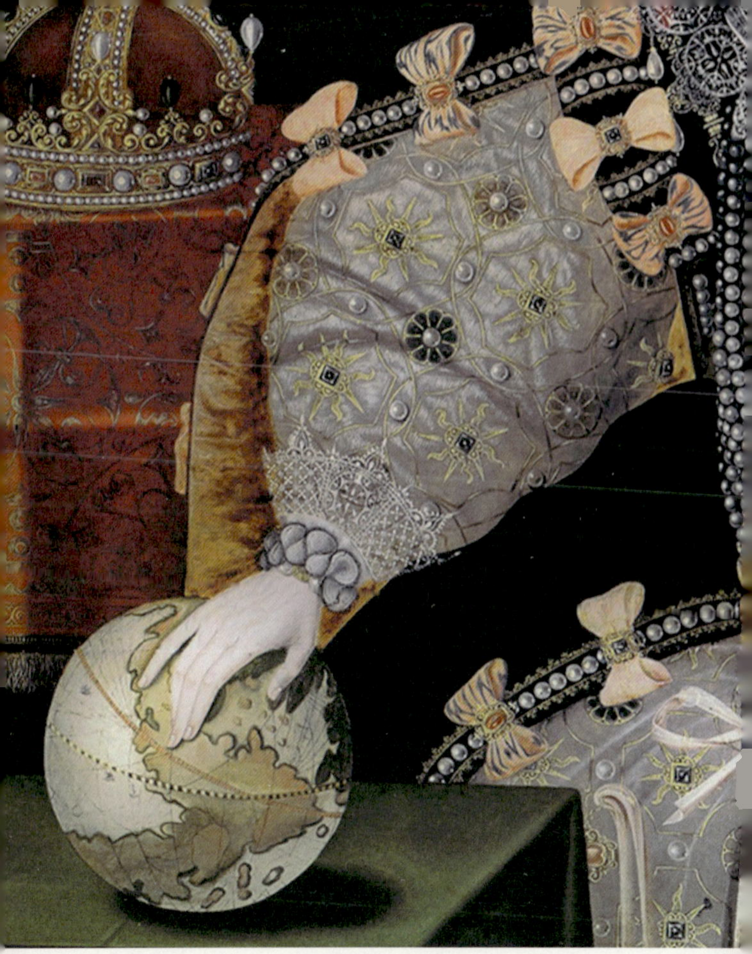

Detail of fresco in the hall dedicated to the science of mathematics in the Uffizi Gallery in Florence, illustrating the principle of the lever from Greek mathematician and astronomer Archimedes who allegedly claimed "Give me a place to stand and I will move the Earth."[6] (courtesy Galleria degli Uffizi)

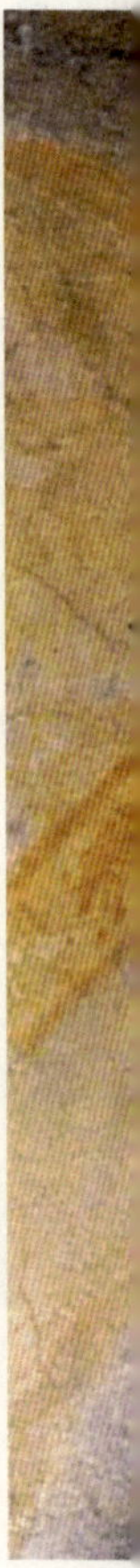

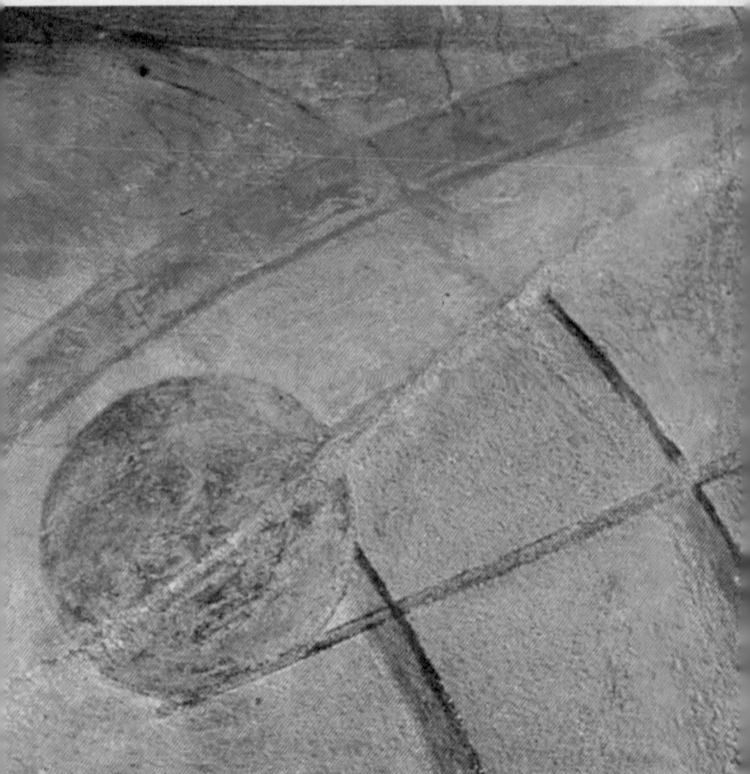

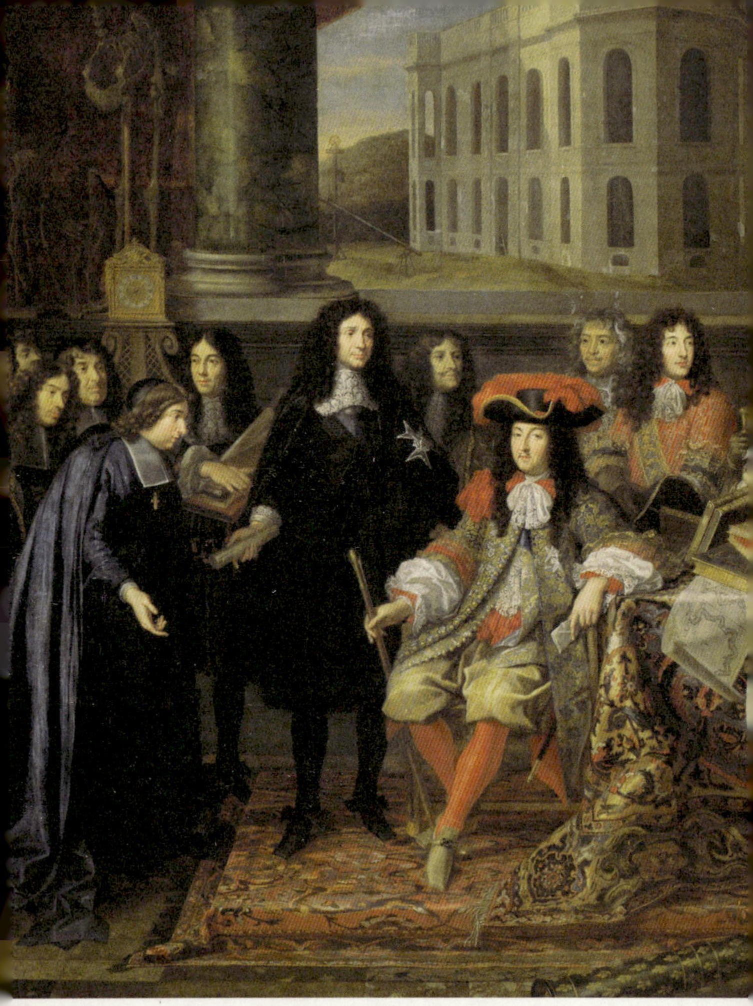

Detail from a Goblin tapestry study by Henri Testelin showing Jean-Baptiste Colbert presenting the members of the newly founded Royal Academy of Sciences to Louis XIV in 1667; the globes and maps depicted in the painting attest to the importance granted to cartographic knowledge in an age marked by the wedding of political power and the sciences.[7] (courtesy Musée national du Château de Versailles)

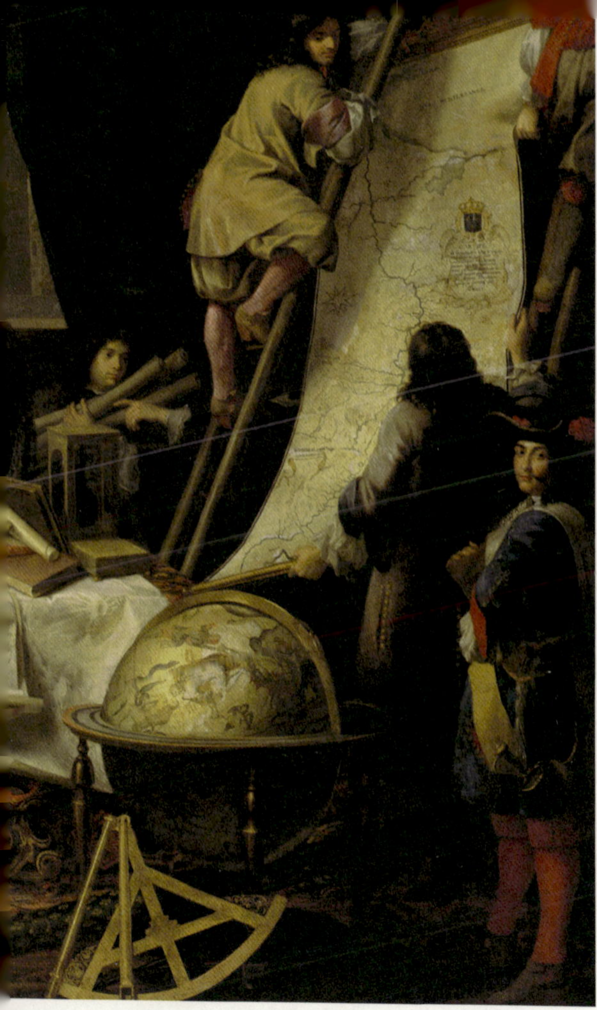

First known representation of Earth's lithosphere by Dutch writer and publisher Willem Goeree from 1694 showing the surface of the globe without water and thus offering a re-visioning of the world that did not resemble the one experienced; this early engraving is strikingly similar to those digitally modelled geoids produced in the field of earth sciences today (courtesy National Archives of the Netherlands / Nationaal Archief)

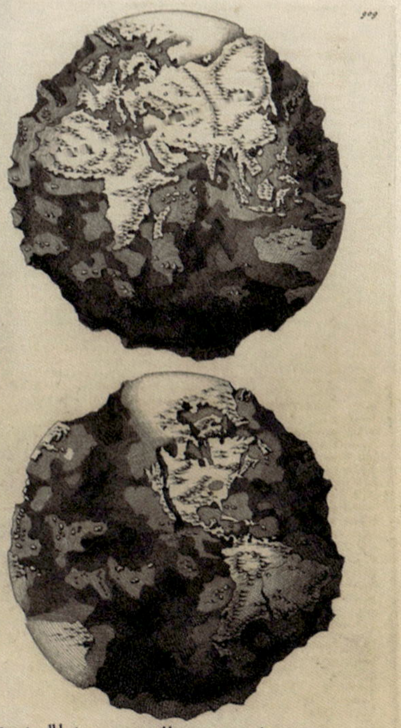

Den Aardkloot van water ontbloot, na twee zyden aante sien.

Drawings by Étienne-Louis Boullée of his Cenotaph for Newton from 1784 showing a large spherical monument to the Newtonian universe, the shell of which is perforated by small star-shaped openings grouped together to form celestial constellations overlooking the sarcophagus inside. (courtesy Bibliothèque nationale de France)

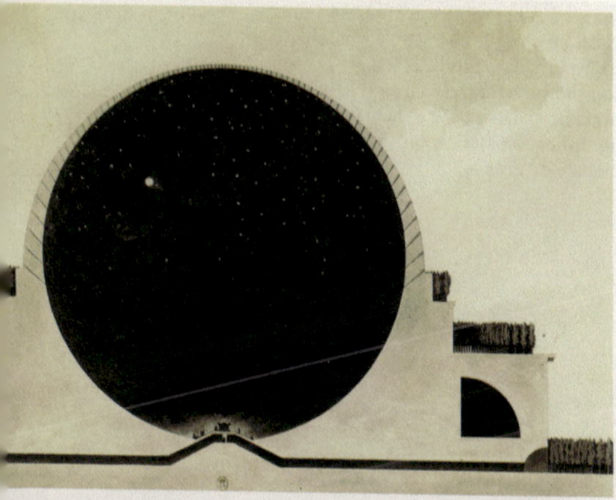

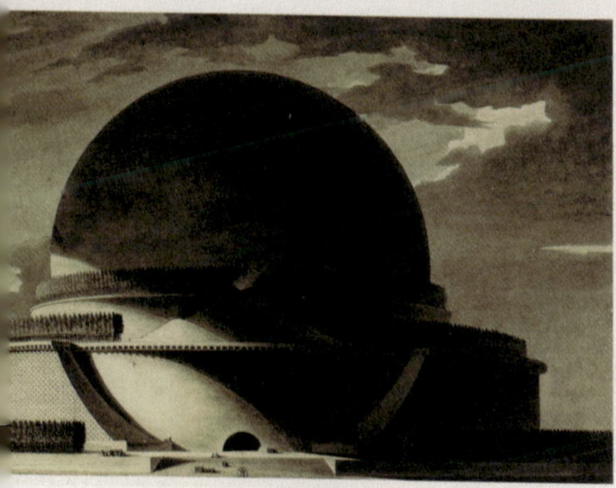

Drawings by Jean-Jacques Lequeu from his 1794 manuscript *Architecture civile* showing an elevation and section of his hypothetical project "Temple of Earth;" the section shows the terrestrial sphere of Earth within a larger celestial sphere punctuated by openings representing the stars. (courtesy Bibliothèque nationale de France)

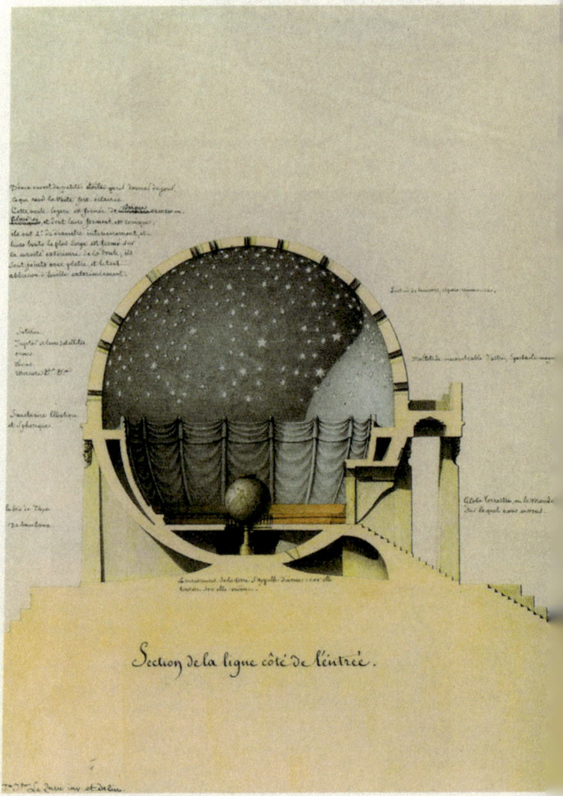

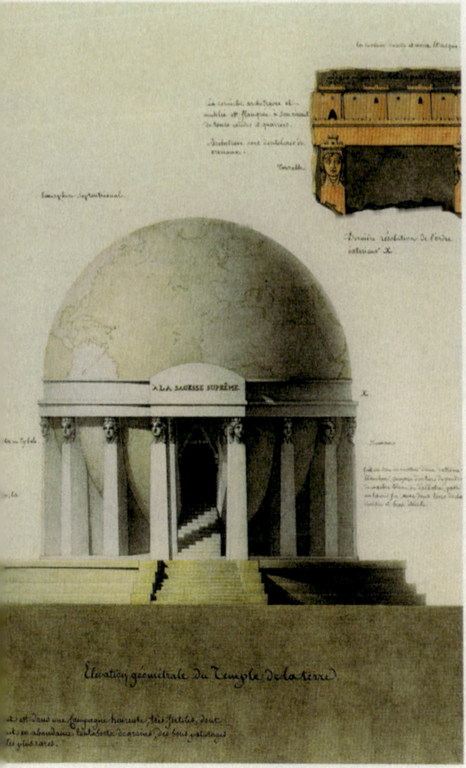
A LA SAGESSE SUPRÊME
Elévation géométrale du Temple de la terre

Satirical, hand-colored etching depicting Napoleon I straddling a globe drawn with little regard to geography to imply a reshuffling of Europe under the Emperor of France, his right foot firmly placed on Switzerland and his left foot resting on Italy; a small figure with a hand on England raises a sword menacingly, to which Napoelon responds "Who is it [that] dares interrupt me in my progress?" (courtesy Trustees of the British Museum)

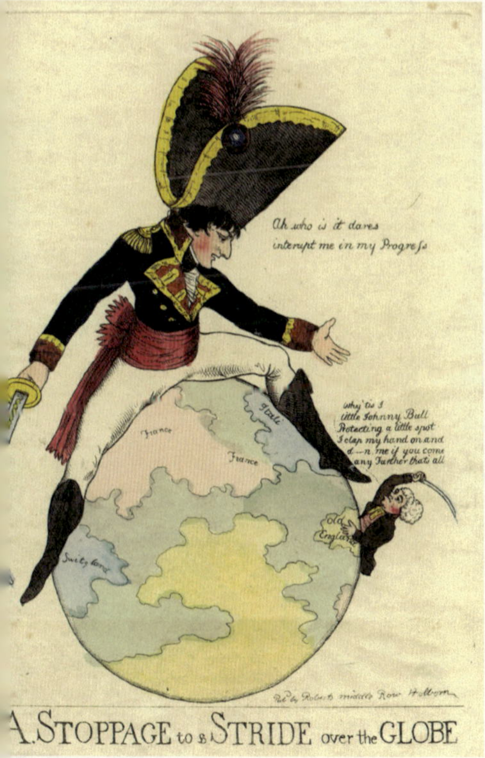
Ah who is it dares
interupt me in my Progress
why 'tis I
little Johnny Bull
Protecting a little spot
I clap my hand on and
d—n me if you come
any Farther that's all
France
France
Itally
Switzerland
Old England
A. STOPPAGE to a STRIDE over the GLOBE

Engraving by Claude-Nicolas Ledoux from his 1804 treatise *L'architecture considérée sous le rapport de l'art, des moeurs et de la législation* showing his theater project at Besançon that, as microcosm of the world, is reflected in the pupil of an all-seeing eye, which presumably is seen as well. (courtesy Bibliothèque Nationale de France)

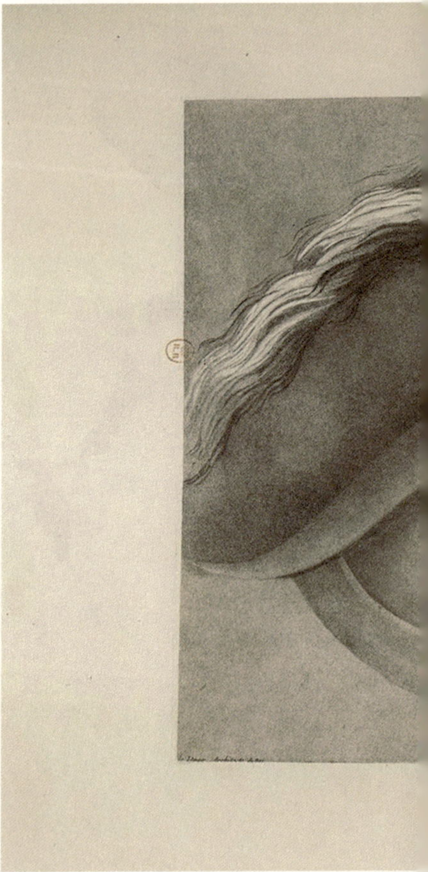

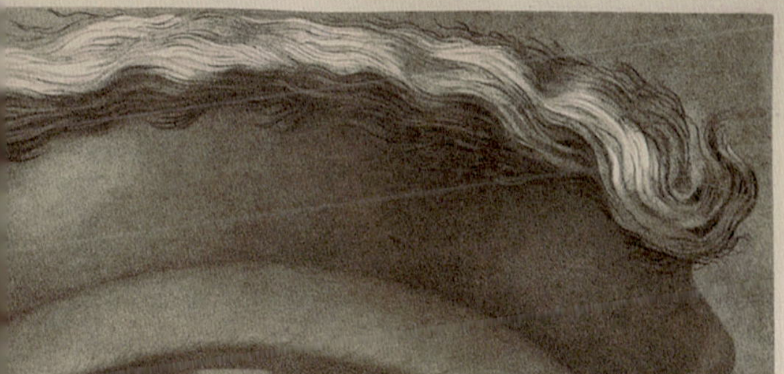

Engraving of a cemetery complex by Claude-Nicolas Ledoux for the town of Chaux from his 1804 treatise *L'architecture considérée sous le rapport de l'art, des moeurs et de la législation*; the central space is a large spherical form that for architects of the revolutionary period represented a meaningful expression eternal in its significance.[8] (courtesy Bibliothèque Nationale de France)

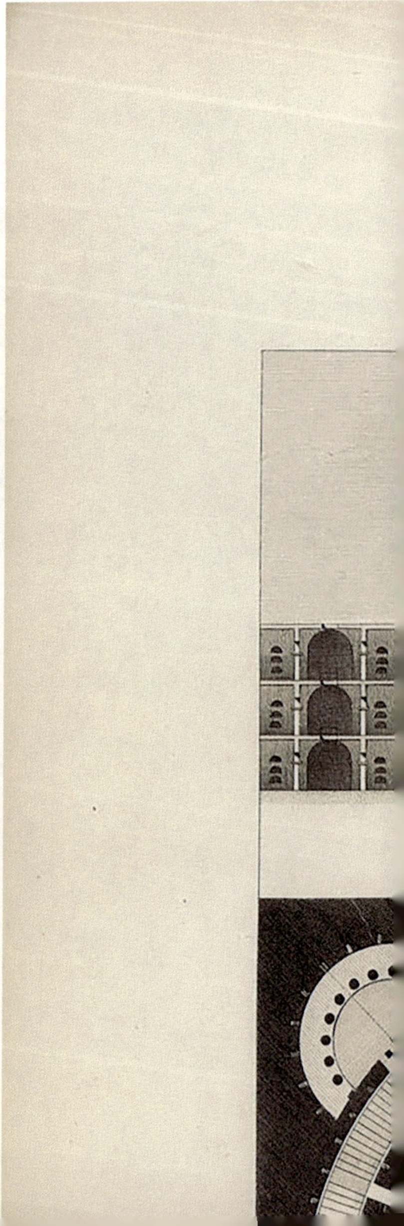

Pl. 99

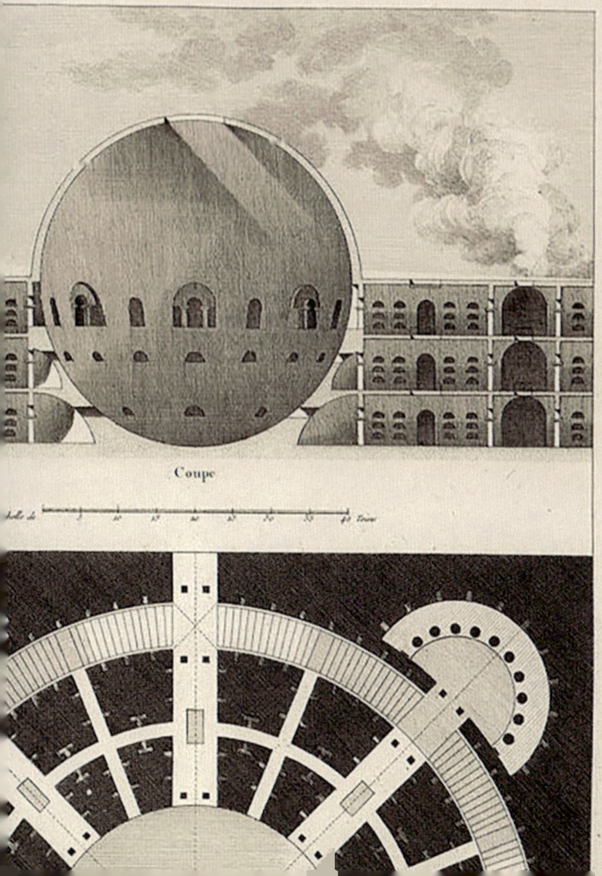

Engraving of the *Monument in Memory of Christopher Columbus* as it appeared on the cover of *Scientific American* in October 1890 announcing the project by architect Alberto Palacio for the Chicago World Fair of 1892; the large statue of Earth, crowned by a replica of Columbus's ship, was to be higher and more spectacular than the Eiffel Tower that had recently appeared as an icon of the Exposition universelle in Paris in 1889; due to excessive construction costs, however, the project was never realized.[9] (courtesy Scientific American)

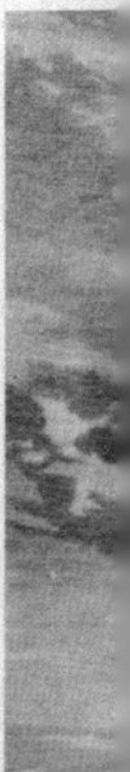

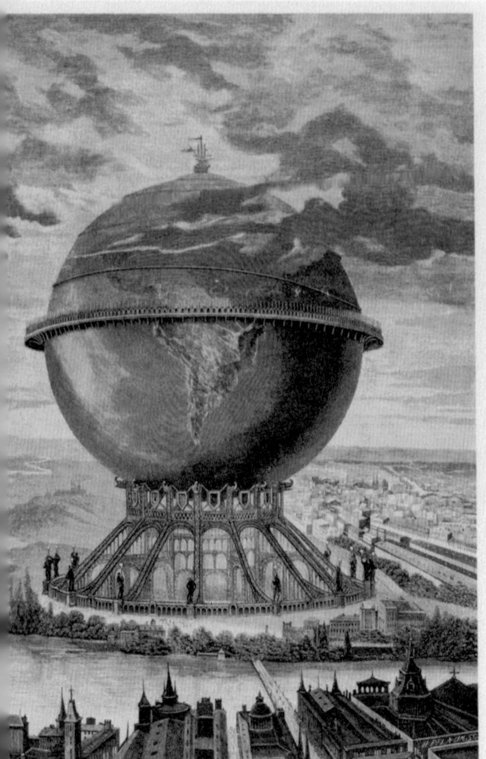

Globe terrestre by geographer Élisée Reclus (drawn by architect Louis Bonnier) for the Exposition universelle of 1900, a showpiece meant to popularize empirical representations of geographical knowledge, if not change prevalent perceptions of the planet generally represented on flat maps; to this end, Reclus presented his 1:100,000 scale Earth to the Sixth International Geographical Congress held in London in 1895, but withdrew his proposal in 1898 due to financial limitations.[10] (courtesy Archives nationales, France)

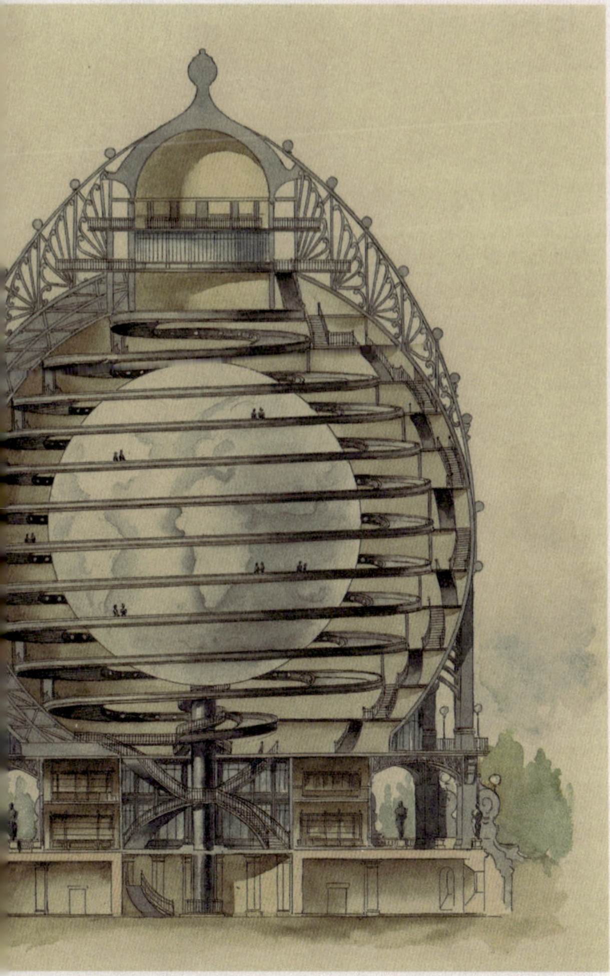

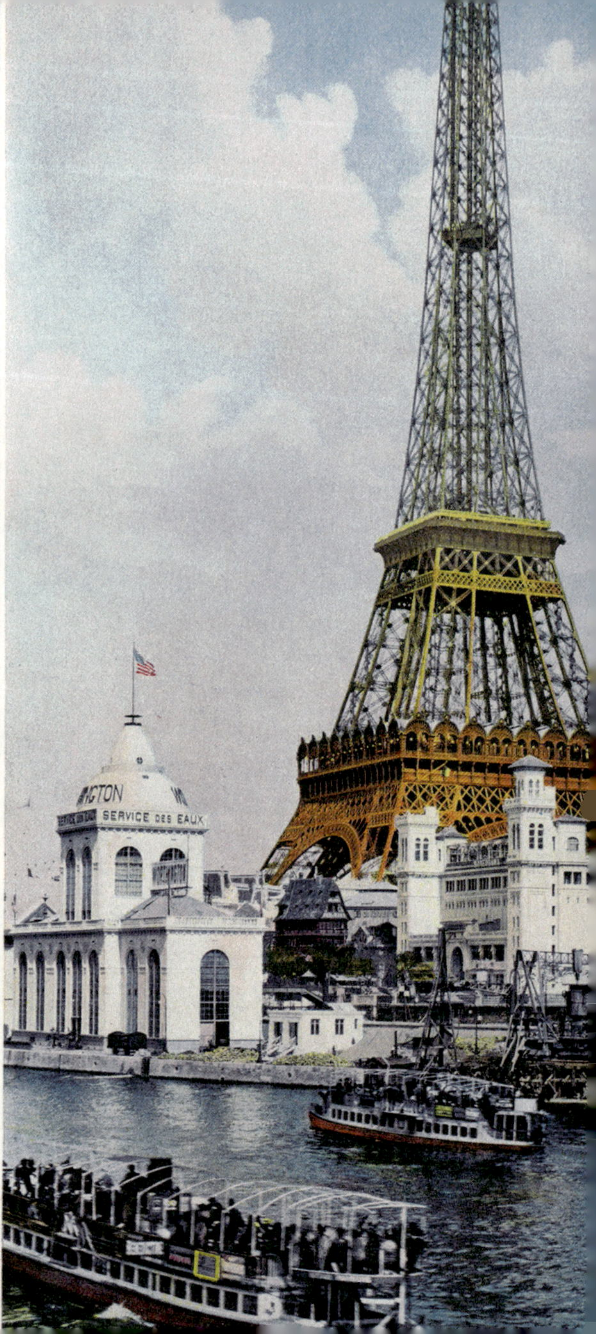
SERVICE DES EAUX

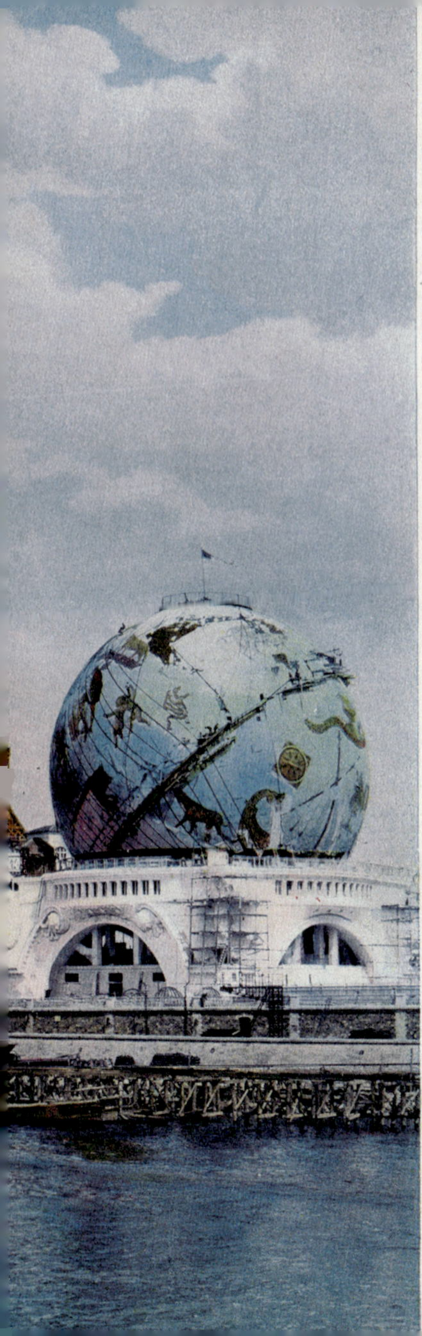

Colored photograph of the *Cosmorama* (or *Globe céleste*) by architects Louis Bonnier and Albert Galeron for the Exposition universelle of 1900, Champs-de-Mars, Paris; the realized structure was a simplified and reduced version of the Élisée Reclus *Globe terrestre* proposed a few years earlier, with a rotating internal globe for visitors encased by a larger sphere on which constellations and zodiac signs were painted; due to structural problems, including the collapse of a pedestrian bridge, the *Cosmorama* was closed to the public during the exhibition.[11] (*Le Figaro illustré*, no. 128, November 1900 and Archives nationales, France)

Photographic portrait of Theodore Roosevelt from January 1, 1903 beside a large globe in the White House during his first term as president; wanting to increase the prestige and influence of the United States on the world stage, he sent the "Great White Fleet" of the US Navy on a world tour to make courtesy calls to various nations, a route that Roosevelt meticulously followed on his globe.[12] (photograph Rockwood Photo Co., courtesy Library of Congress)

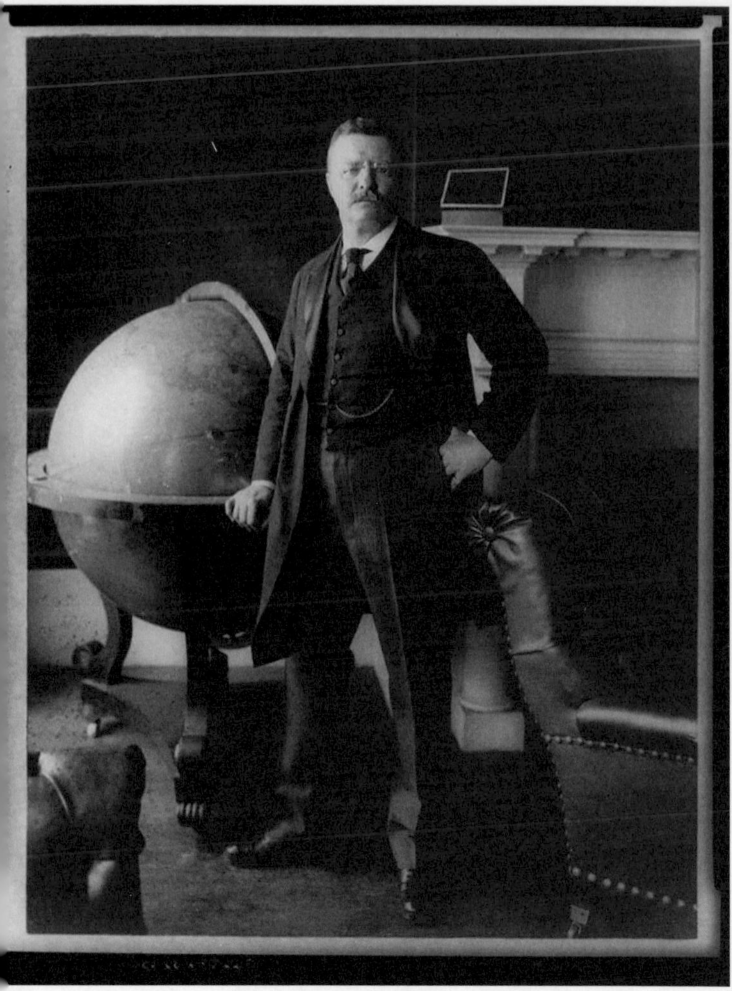

Political satire by Viktor Deni entitled "Comrade Lenin cleans the Earth from filth" showing Vladimir Lenin with red broom in hand ridding the world of the aristocracy, the clergy, and the bourgeoisie; the poster was printed by the state publishing house Litizdat, where Viktor Deni worked at the time, November 1920. (courtesy cultural-images)

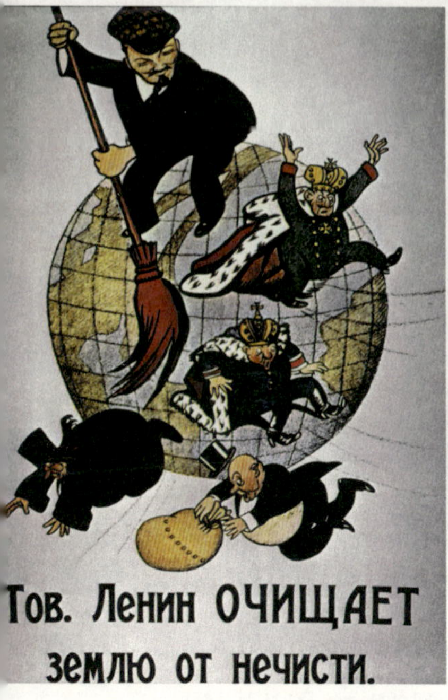
Тов. Ленин ОЧИЩАЕТ
землю от нечисти.

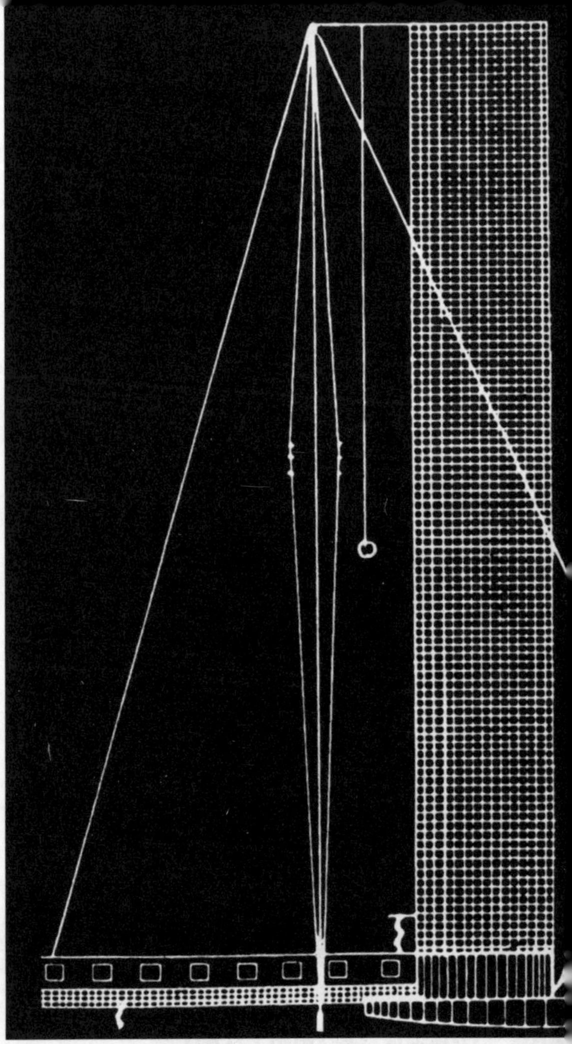

1927 thesis project by Ivan Illich Leonidov for the Lenin Institute for Librarianship (the collective scientific and cultural center of the USSR) at the main art and technical School of Moscow (VKhUTEMAS), under the supervision of Alexander Vesnin; the project is comprised of Platonic forms: a spherical auditorium that also functions as a planetarium and a tower housing the library.[13] (photograph A. Kopp, see Catherine Cooke, *Architectural Drawings of the Russian Avant-Garde*, The Museum of Modern Art, New York, 1990)

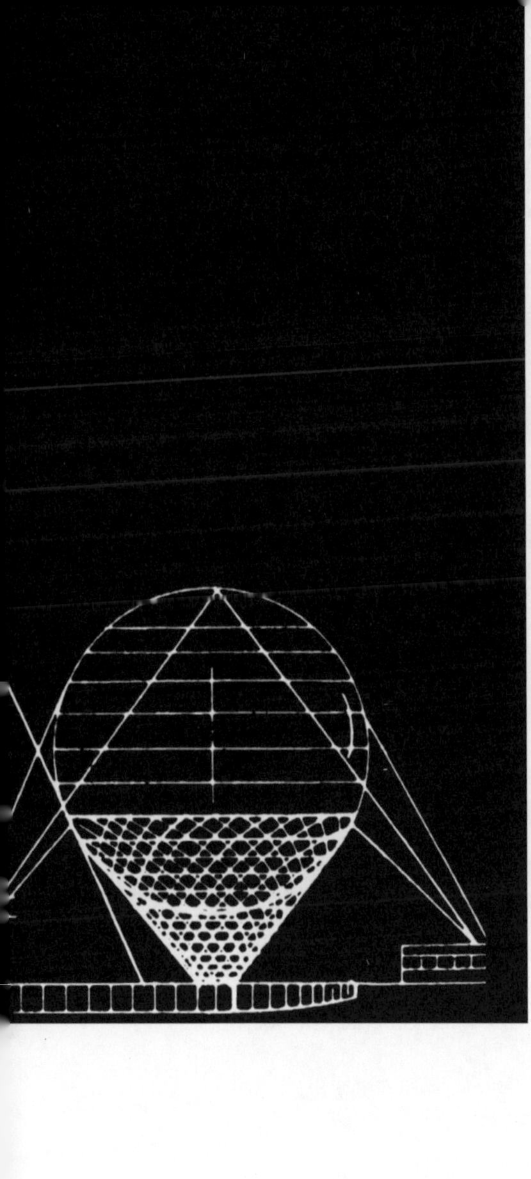

While on a retreat in 1927, Buckminster Fuller sketched out his vision of a "4D Interconnected, Unified World," also known the "4D World Town Plan," which speculated on the implications of an entirely urbanized planet, claiming that 2 billion new homes would be required by the end of the century.[14] (courtesy Buckminster Fuller Institute)

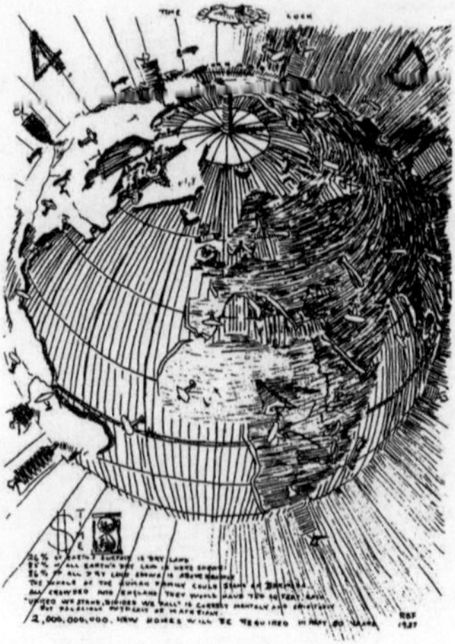
TIME
26% OF EARTH'S SURFACE IS DRY LAND
2,000,000,000. NEW HOMES WILL BE REQUIRED IN NEXT 50 YEARS
RBF

1934 drawing by Buckminster Fuller of a "one-world-island in a one-world-ocean," which was the basis for his Dymaxion World Map from 1944 that could be unfolded and rearranged in a multiplicity of ways, allowing different world regions to occupy the center of the map; the new projection method was patented in 1946.[15] (courtesy Buckminster Fuller Institute)

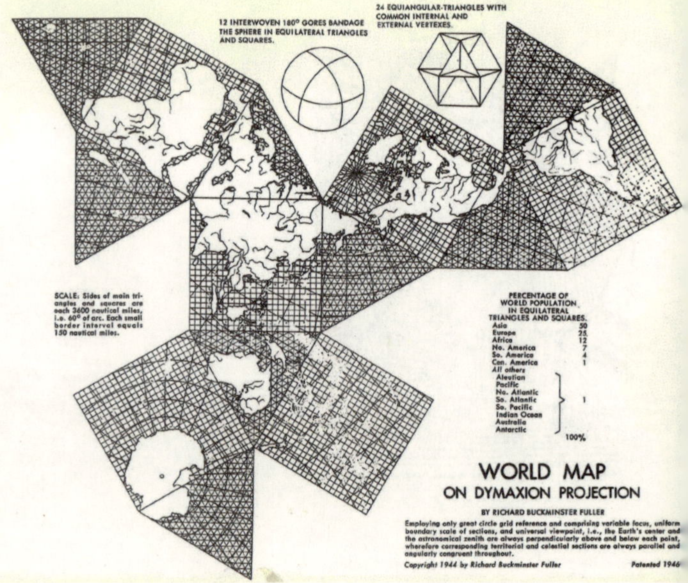

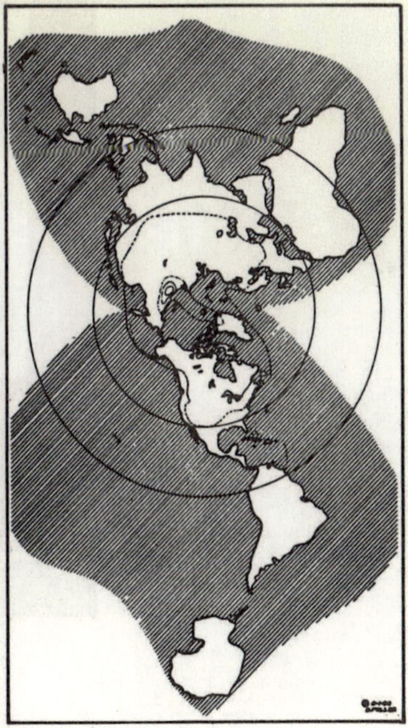

Constructed for the 1939 New York World's Fair, the Perisphere, in keeping with the fair's theme "The World of Tomorrow," housed a diorama called "Democracity" that displayed a utopian city of the future; the spherical structure was accompanied by a tall spire-shaped Trylon, both of which were designed by architects Wallace Harrison and André Fouilhoux.[16] (photograph Leo Husick, courtesy Library of Congress)

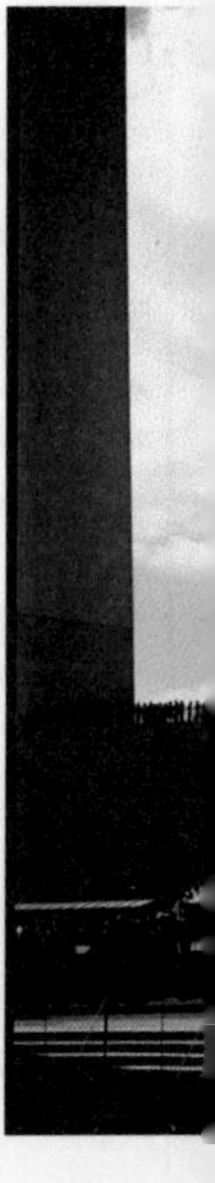

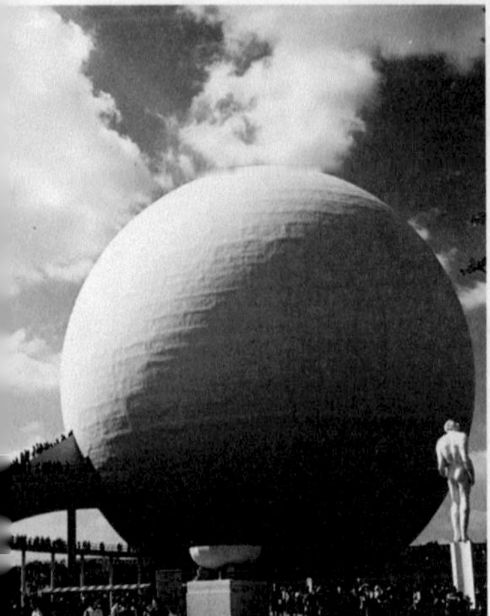

Adolf Hitler in the Reichskanzlei on March 11, 1939 receiving Japanese ambassadors; in the background is the Columbus Globe for State and Industry Leaders or "Führer globe," the only two versions of which were produced for the Nazi party and Hitler in the mid-1930s. (courtesy akg-images)

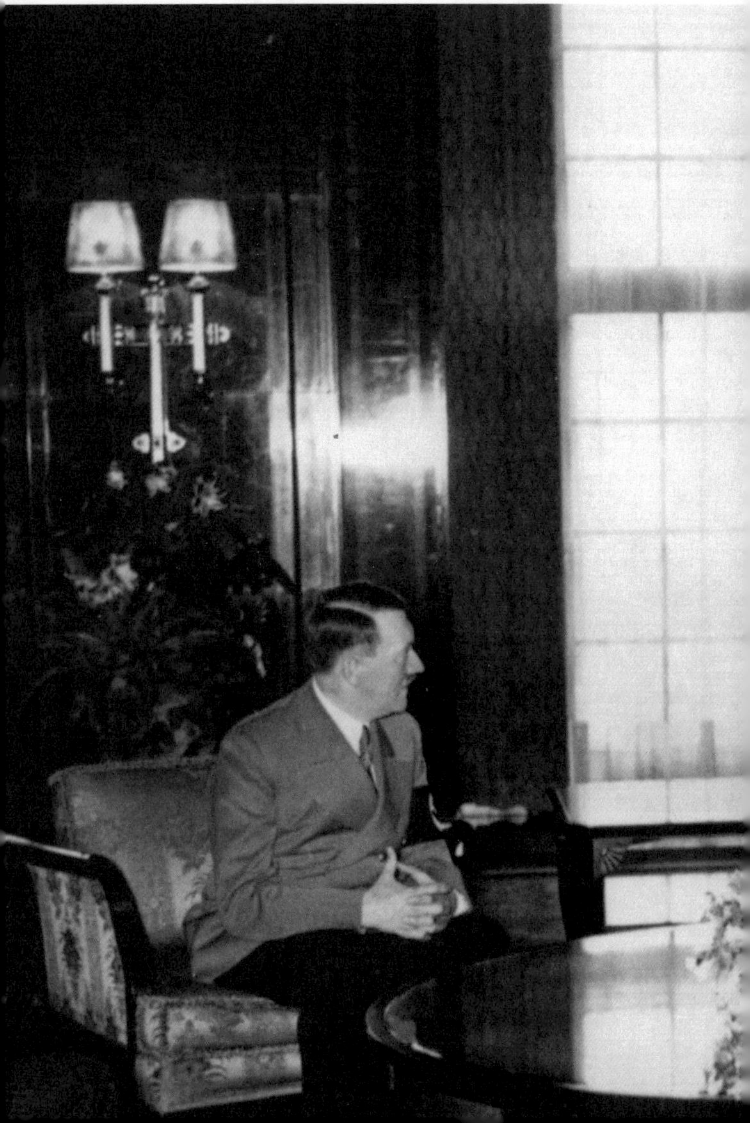

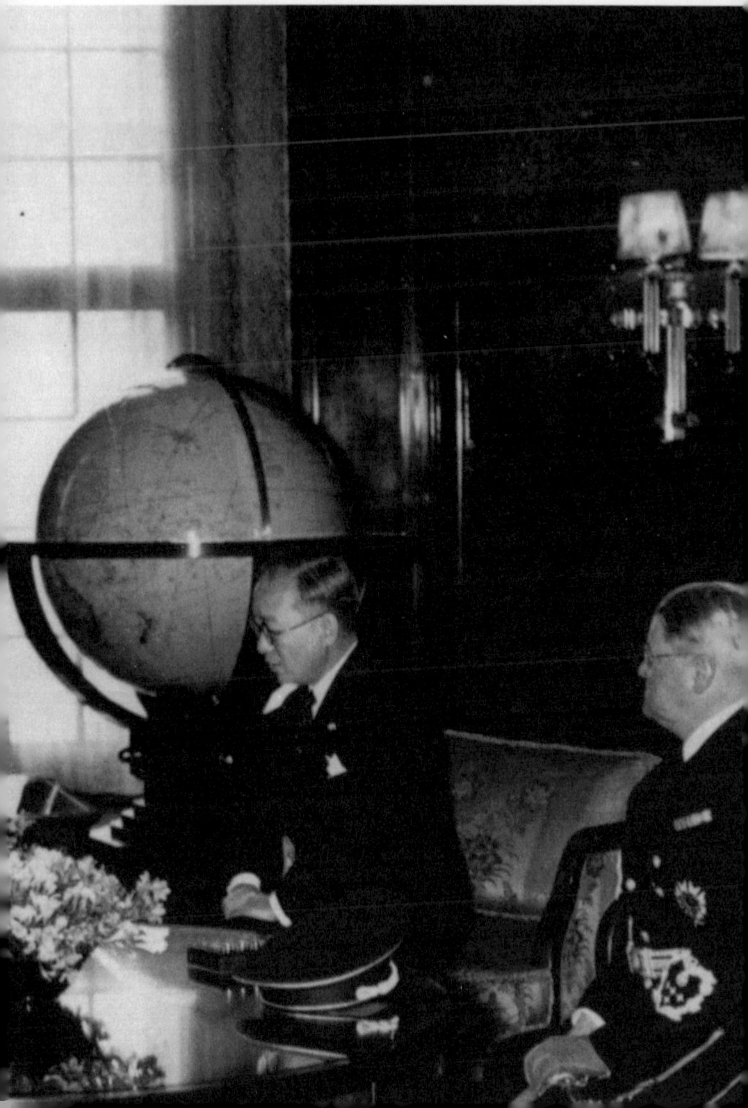

Still from Charlie Chaplin's 1940 satirical film parodying Hitler and the rise of authoritarian regimes entitled *The Great Dictator*, in which the main character is seen performing – to Richard Wagner's opera *Lohengrin* (a favorite of Hitler) – a pas-de-deux with a balloon globe that, believed to be within his grasp, ultimately bursts. (courtesy Chaplin Foundation, ©1940 Roy Export S.A.S., renewed ©1968 Roy Export S.A.S.)

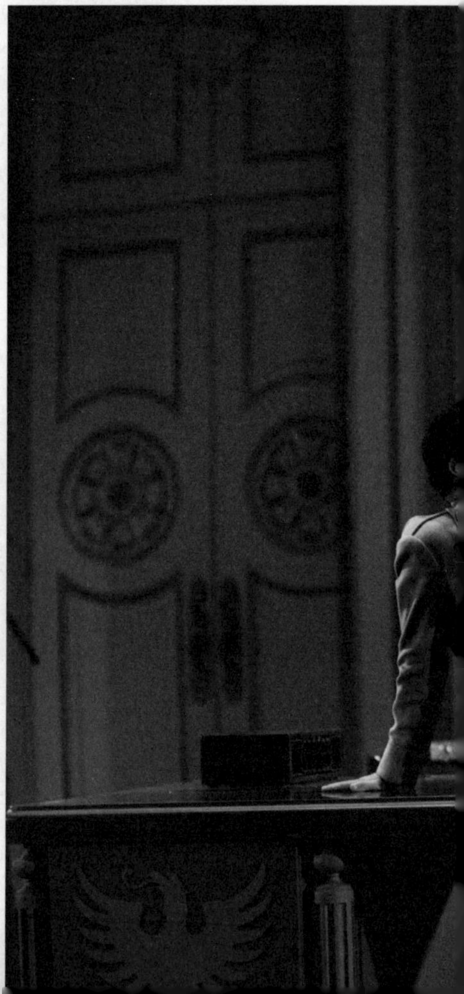

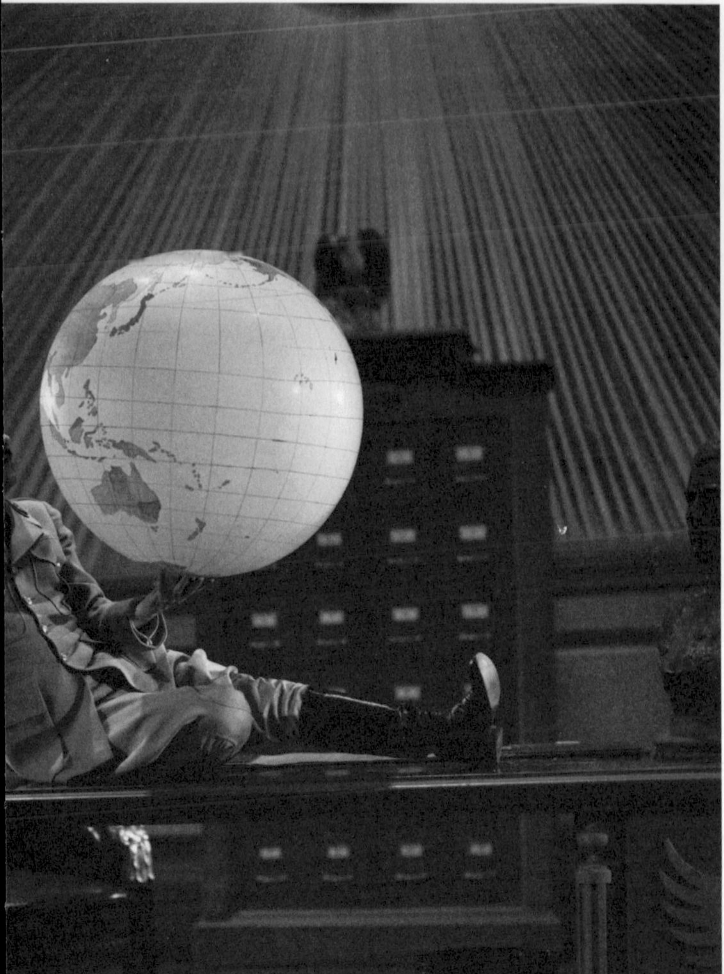

Photograph of Franklin D. Roosevelt in the White House admiring the so-called "President's Globe," which was presented to him as a Christmas gift by the United States Army in December 1942, a globe designed specifically for his consultation during World War II; matching globes were produced for British Prime Minister Winston Churchill and the United States Chief of Staff George C. Marshall. (courtesy Franklin D. Roosevelt Presidential Library and Library of Congress)

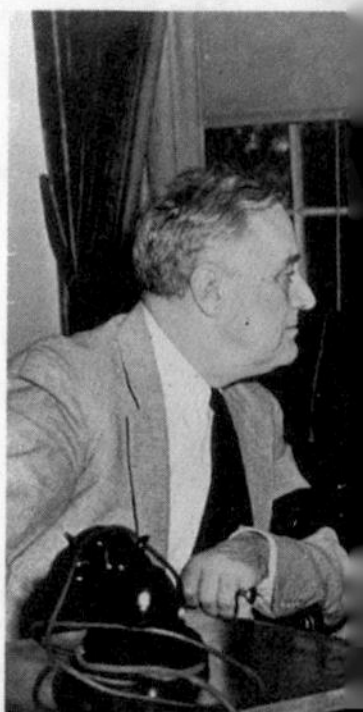

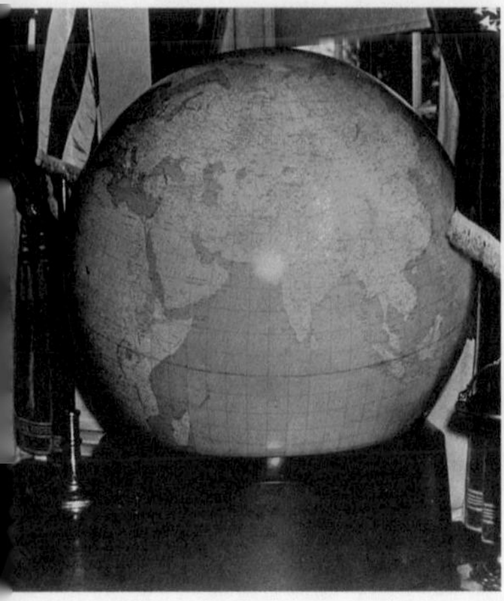

Photograph showing Russian Red Army soldiers posing around Adolf Hitler's Columbus Globe for State and Industry Leaders or "Führer Globe" in the Reich Chancellery in Berlin, May 1945. (photograph G. Kapustyansky, courtesy Russisches Staatsarchiv für Film- und Fotodokumente)

[illegible]

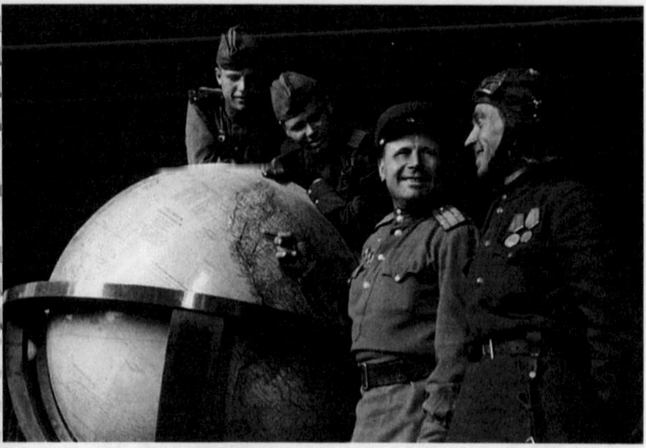

Original sketch of the emblem of the United Nations by graphic designer Donal McLaughlin under the supervision of architect Oliver Lincoln Lundquist in preparation for the first UN Conference on International Organization in 1945, when the UN Charter was drafted; the sketch shows a top-down view of the globe with North America centrally located, embraced by olive branches to symbolize world peace; the later adopted logo is a slightly modified version of this design, with the Unites States rotated to the left. (courtesy United Nations and Donal McLaughlin)

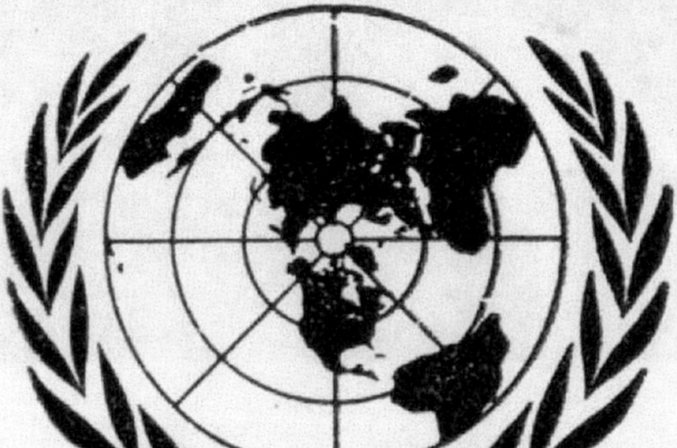

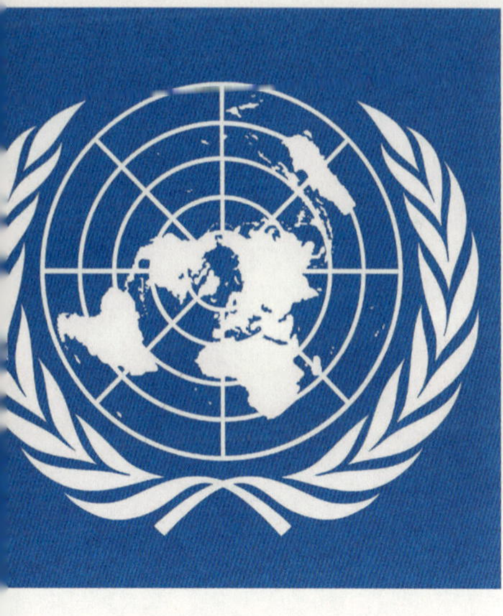

Inflated domes of the so-called Pentadome at Andrews Air Force Base in Maryland that is used as a missile maintenance hangar; the structure, made of nylon skins held up by air pressurc from blowers at its base, was designed by architect Kenneth E. Joy and engineer Walter Bird, 1958-59; the central dome has the Northern Hemisphere printed on its exterior.[17] (courtesy United States Army)

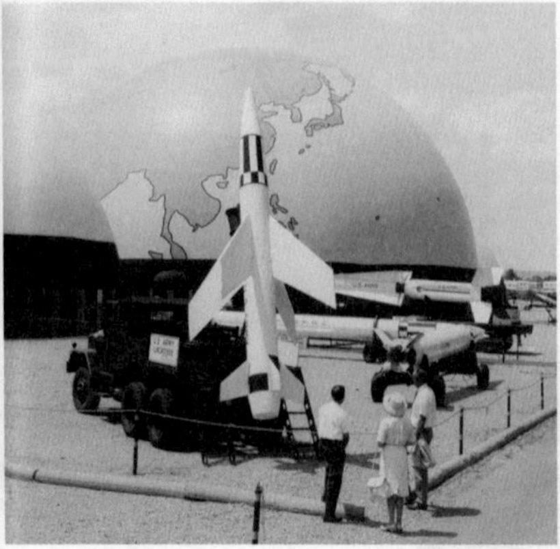

Photograph of a photograph, the latter entitled *Yves Klein et le Globe terrestre bleu* (Yves Klein and the blue globe), April 12, 1961: "In 1957, Yves Klein stated that the Earth was entirely blue. It was on that occasion that he created a Blue Relief globe. Four years later, cosmonaut Gagarin stated in April 1961 that the Earth is of a deep intense blue!!!" (quote from *Yves Klein le Monochrome, il nuovo realismo del colore*, Milan, Galleria Apollinaire, November 1961; original photograph Harry Shunk and Janos Kender, courtesy The Getty Research Institute, Los Angeles; photograph of photograph Hans Frei, 2019)

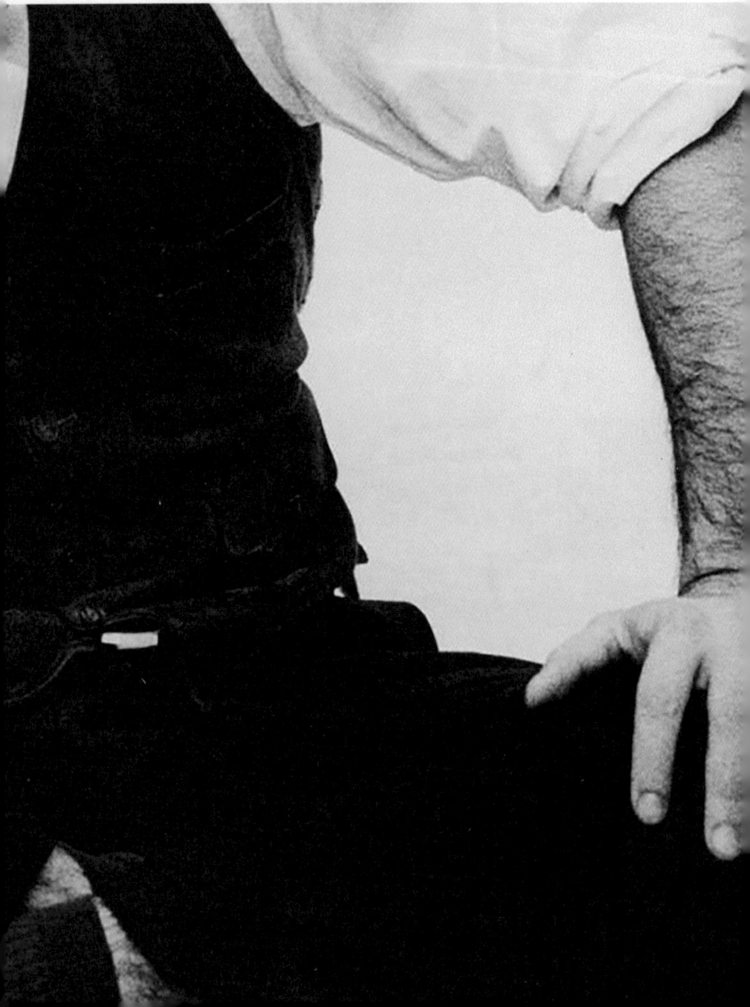

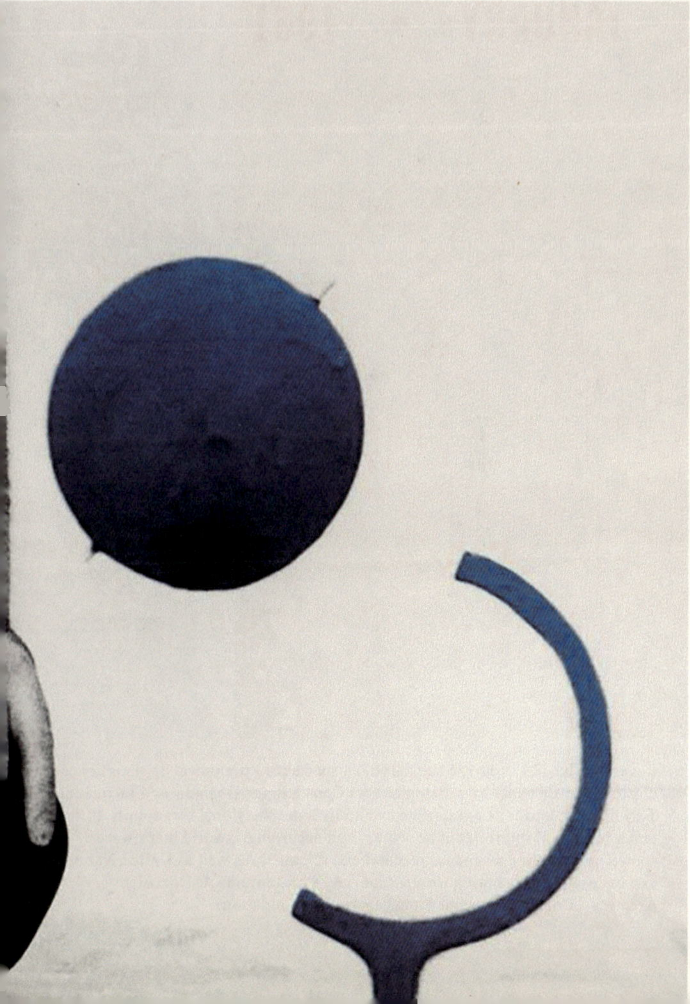

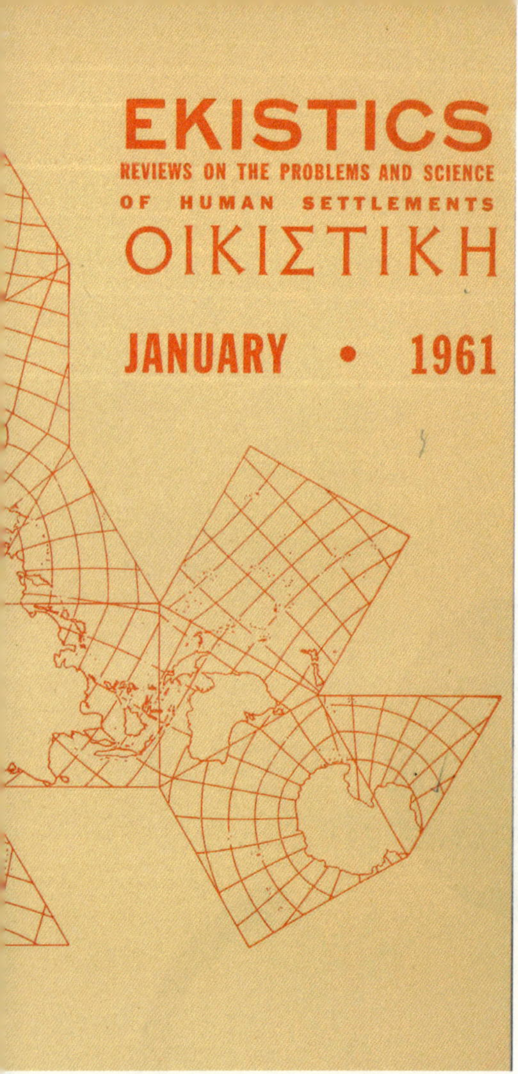

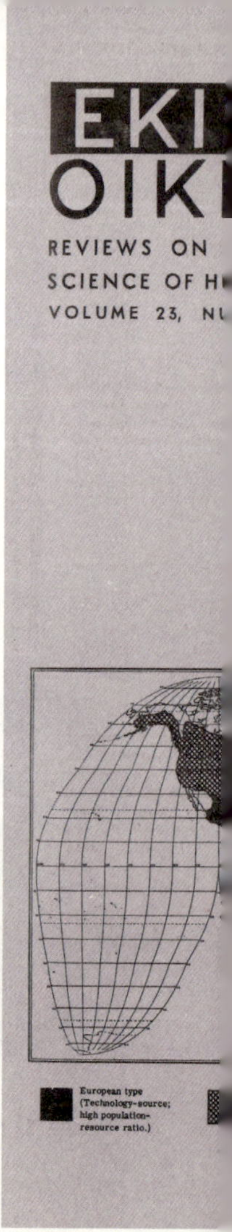

Covers of *Ekistics* from 1961 and 1967, a specialized periodical on the science of human settlements by urban planners Constantinos Doxiadis and Jacqueline Tyrwhitt; the journal's mission was to understand the global community as an entity formed of interrelated networks, and accordingly, world maps were shown on the cover to update researchers, scientists, as well as United Nations experts on global planning issues. (courtesy Constantinos A. Doxiadis Archives / Constantinos and Emma Doxiadis Foundation)

TICS
TIKH
ROBLEMS AND
SETTLEMENTS
138, MAY 1967

STRATEGY FOR DEVELOPMENT

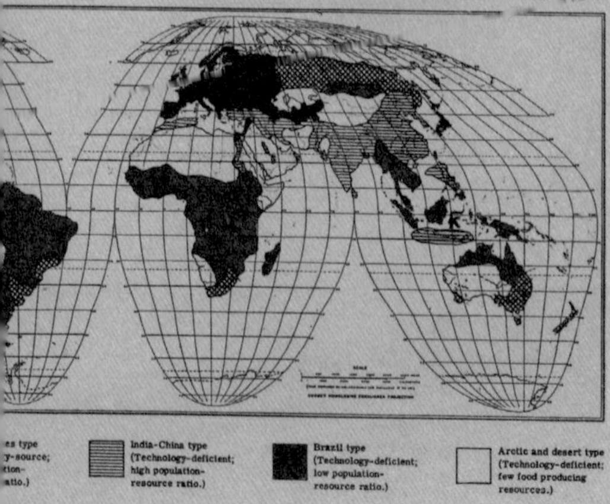

Ecumenopolis, a term introduced by Constantinos Doxiadis as early as 1961 to describe urbanization patterns that would eventually span the globe, was coined to address the controversial concept of a 'world city' that would saturate the available habitable land of Earth by the end of the 21st century; the two global maps shown are scenarios for the end of the 20th and the end of the 21st centuries, respectively.[18] (courtesy Constantinos A. Doxiadis Archives / Constantinos and Emma Doxiadis Foundation)

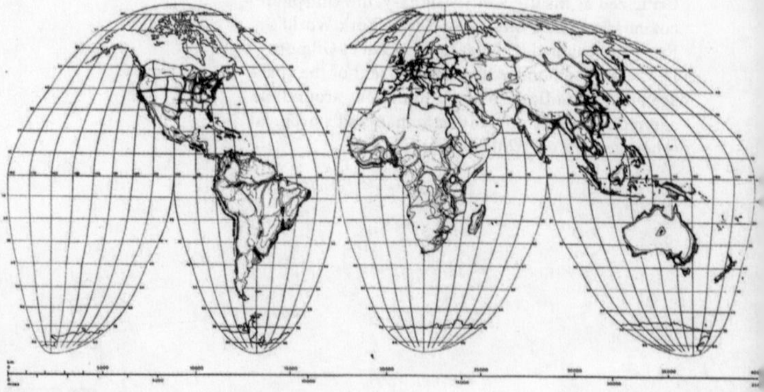

Densities:

- high
- medium
- low
- deep ocean waters

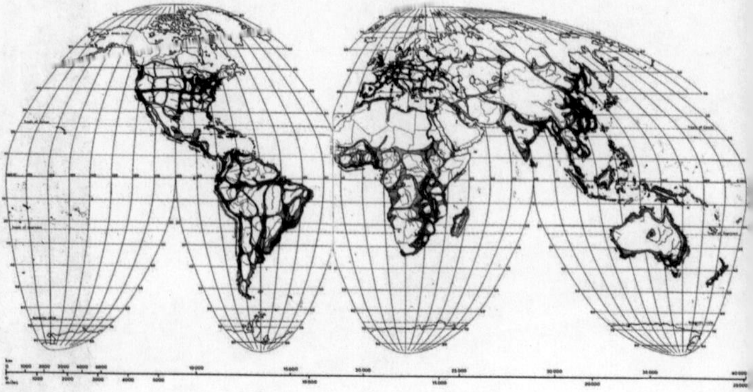

Densities:

- high
- medium
- low

Criticized as insufficiently visionary, the Unisphere – commissioned for the 1964-65 New York World's Fair and designed by landscape architect Gilmore D. Clarke – symbolized both the advent of the space age and the unification of world nations around the common cause of celebrating humankind's progress. (photograph Ron White, 1965)

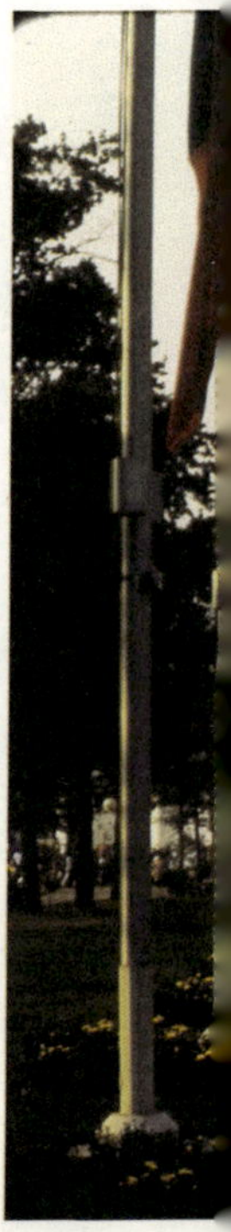

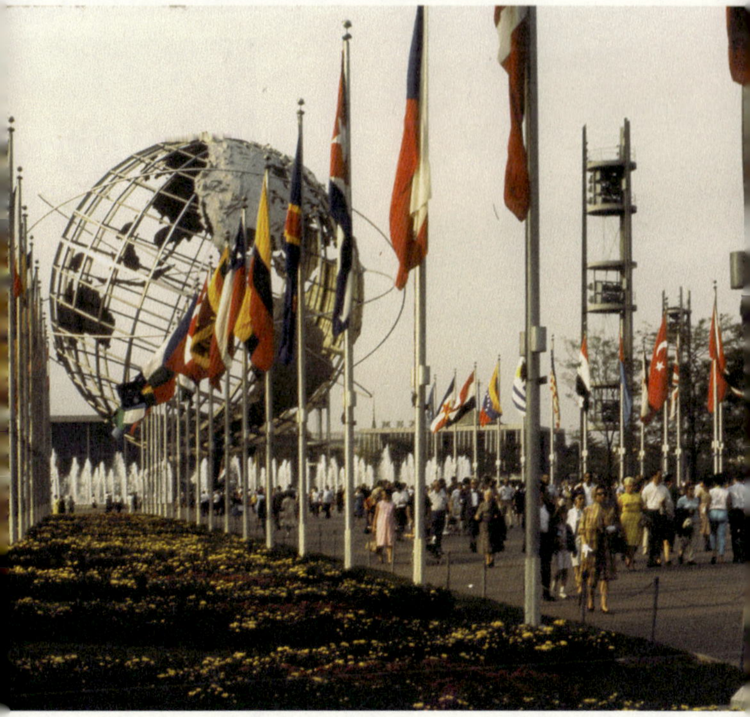

Having spent nearly 20 years developing his geodesic domes, architect Buckminster Fuller was commissioned for the design of the Montreal Biosphere, the United States pavilion for the 1967 World Exposition; the structure's envelope burned in 1976 and the facility was subsequently rebuilt to house the Biosphere Environment Museum that opened in 1995.[19] (photograph Magnum Photos)

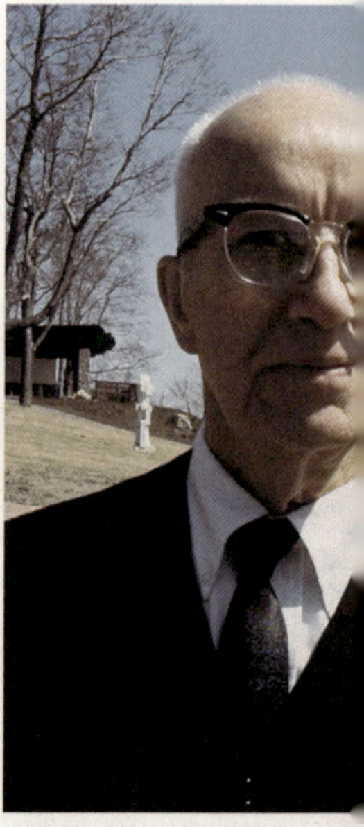

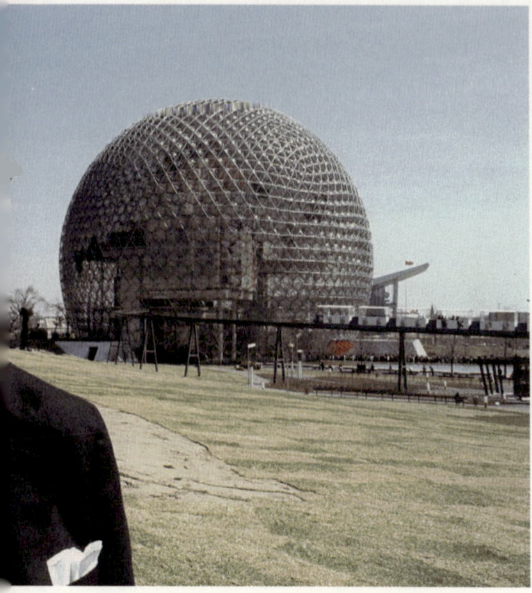

First color image of Earth, a composite of photographs taken in 1967 by the Applications Technology Satellite (ATS-3) of the National Aeronautics and Space Administration (NASA) that was launched into geosynchronous orbit at an altitude of 35,800 km above Brazil on November 5, 1967; the image was used for the cover of the first edition of the *Whole Earth Catalog: Access to Tools* in the fall of 1968. (courtesy NASA)

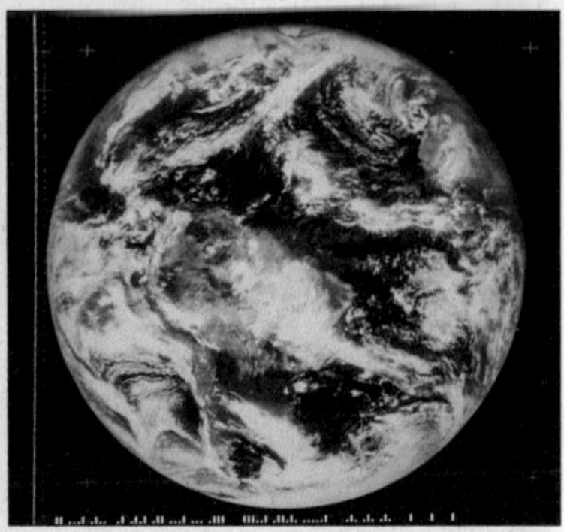

"We came to explore the Moon and what we discovered was the Earth," claimed lunar module pilot William Anders of the Apollo 8 mission after having taken the iconic photograph *Earthrise* on December 24, 1968, at the time the most reproduced photograph; often published by NASA with Earth at the top, the original photograph had Earth on the left.[20] (courtesy National Aeronautics and Space Administration, NASA)

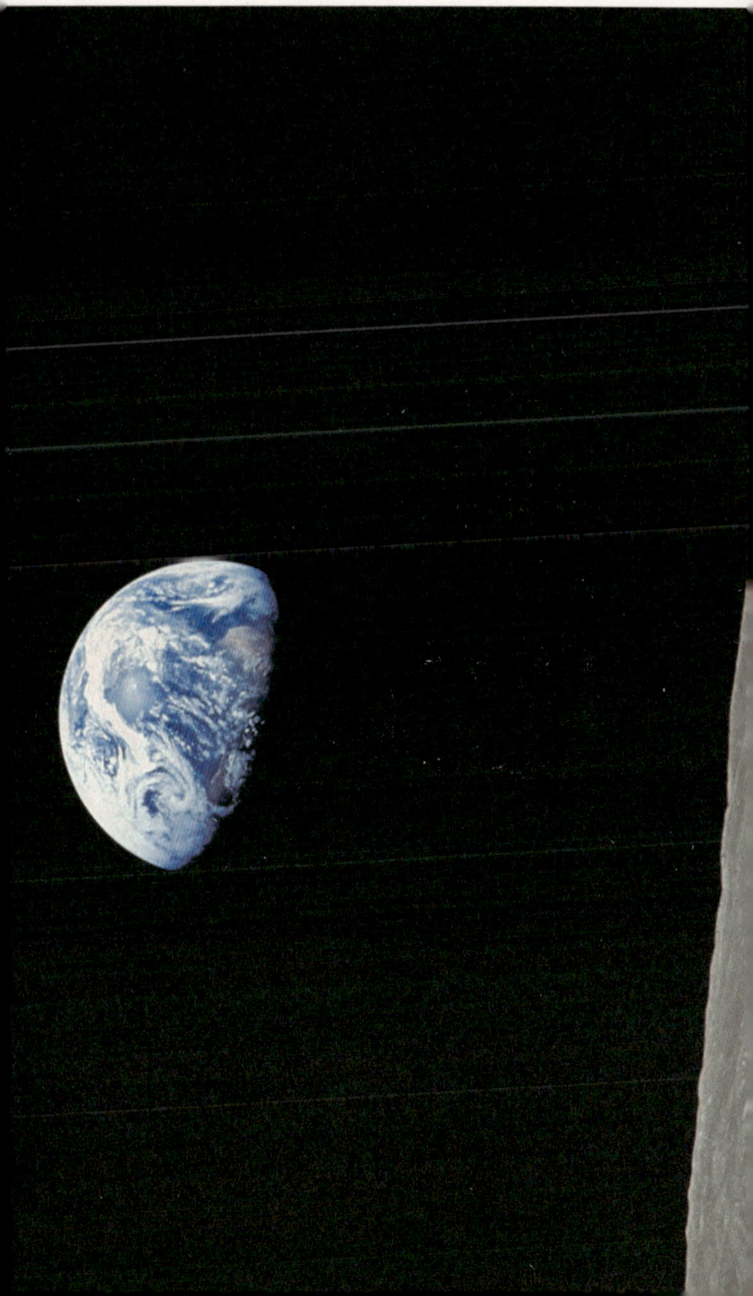

In 1968, Los Angeles architects Ray and Charles Eames produced what they called *A Rough Sketch for a Proposed Film Dealing with the Powers of Ten and the Relative Size of Things in the Universe*, in which they explored relations of scale from the atom to the edge of the known universe; this sketch would lead to the more known film *The Powers of Ten*, produced in 1977 and sponsored by IBM. (courtesy Eames Office)

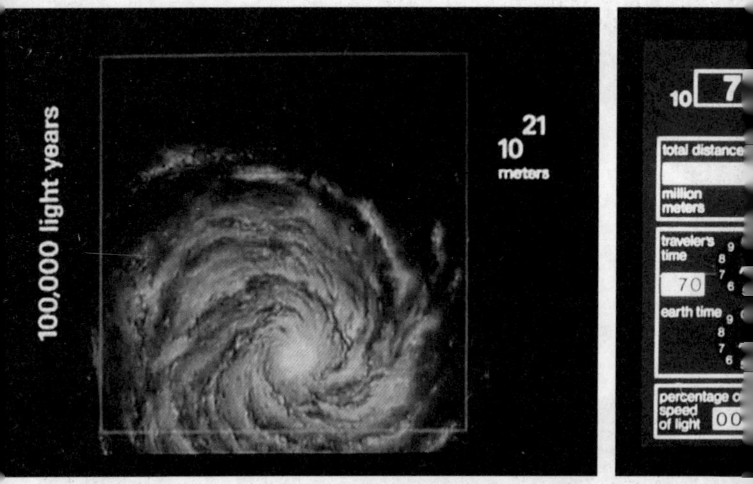

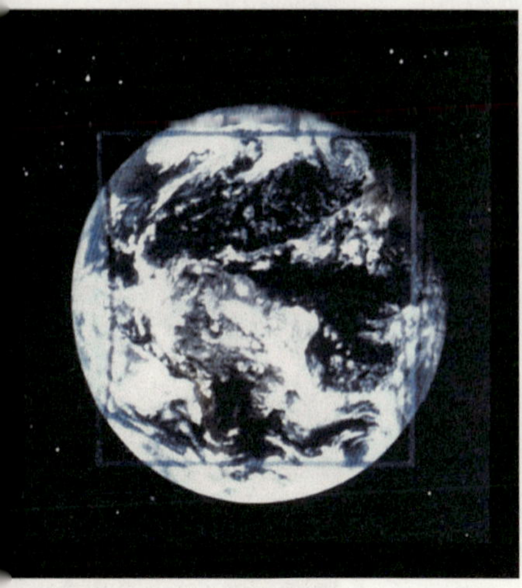

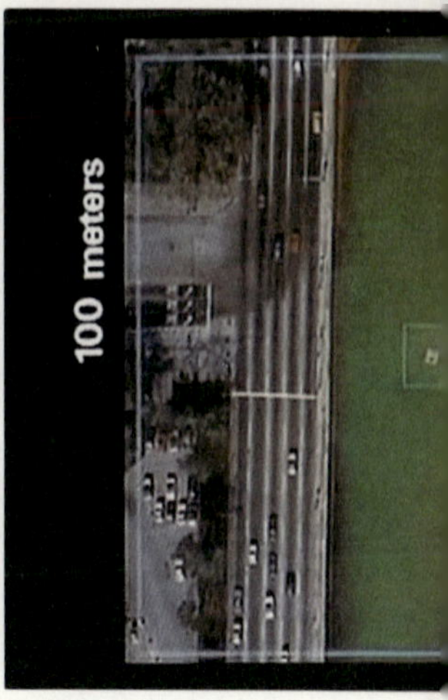
100 meters

First conceived as part of the core curriculum of the Southern Illinois University Edwardsville in 1961, the *World Game* by Buckminster Fuller was to be deployed as a tool to facilitate the cooperative development of ecologically-sound solutions for the future of the world; it was first presented to a larger audience at the 1967 Universal Expo in Montreal and was played for the first time at the New York Studio School for Drawing, Painting, and Sculpture in 1969. During the following decade, the game evolved and expanded through seminars, lectures, publications, and exhibitions, resulting in the founding of the World Game Institute in Philadelphia in 1972.[21] (courtesy Buckminster Fuller Institute)

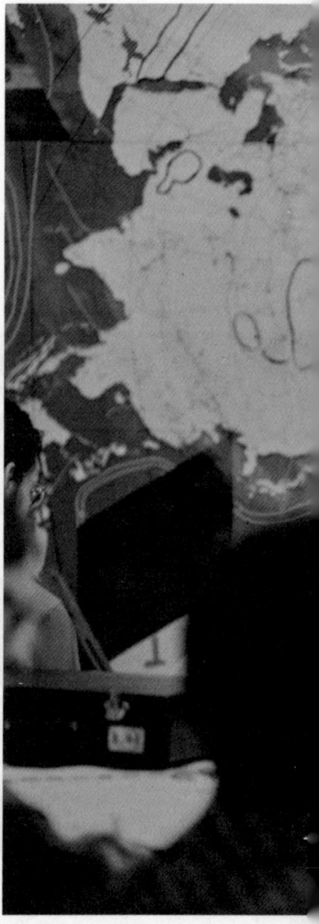

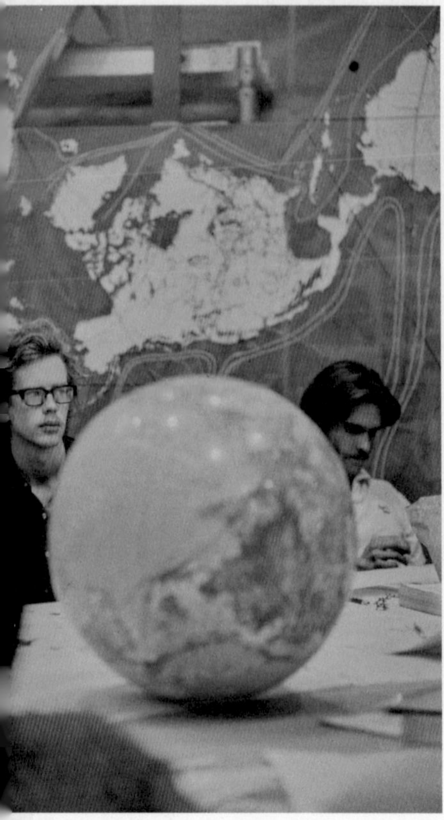

Robert Leydenfrost's poster portraying the globe wearing a gas mask that announced the first Earth Day on April 22, 1970, which, among other activist responses to reckless human exploitation of nature, prompted the rise of the environmental movement at the time, primarily in the United States and soon thereafter in Europe.[22] (courtesy Robert Leydenfrost)

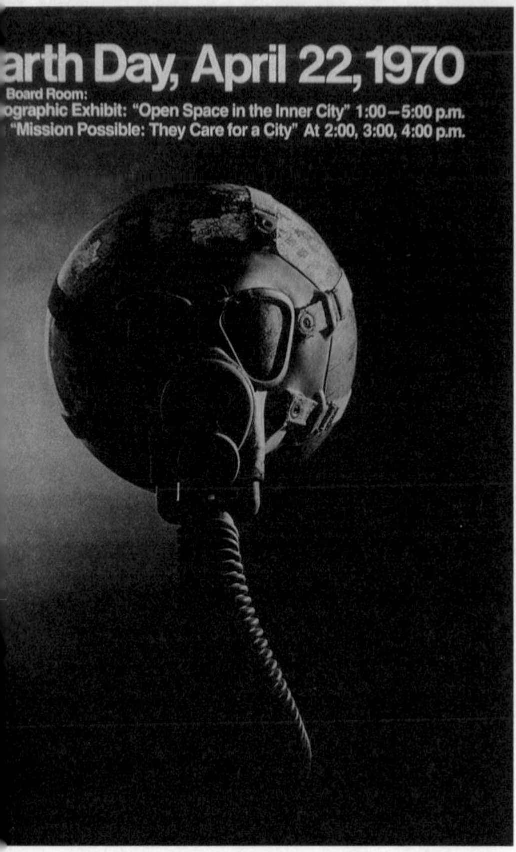
arth Day, April 22, 1970
Board Room:
ographic Exhibit: "Open Space in the Inner City" 1:00—5:00 p.m.
"Mission Possible: They Care for a City" At 2:00, 3:00, 4:00 p.m.

Images of the front and back cover of the *The Last Whole Earth Catalog* from 1971, for which its publisher Stewart Brand received the National Book Award in the following year; the photographs chosen for this edition were actually a composite of images taken by the ATS-3 satellite, which offered the first view of the 'whole Earth' and signaled the onset of a new worldview.[23] (submitted by Stephen Cairns, Future Cities Laboratory Singapore, Department of Architecture, ETH Zurich, courtesy Stewart Brand and the Department of Special Collections and University Archives, Stanford University Libraries)

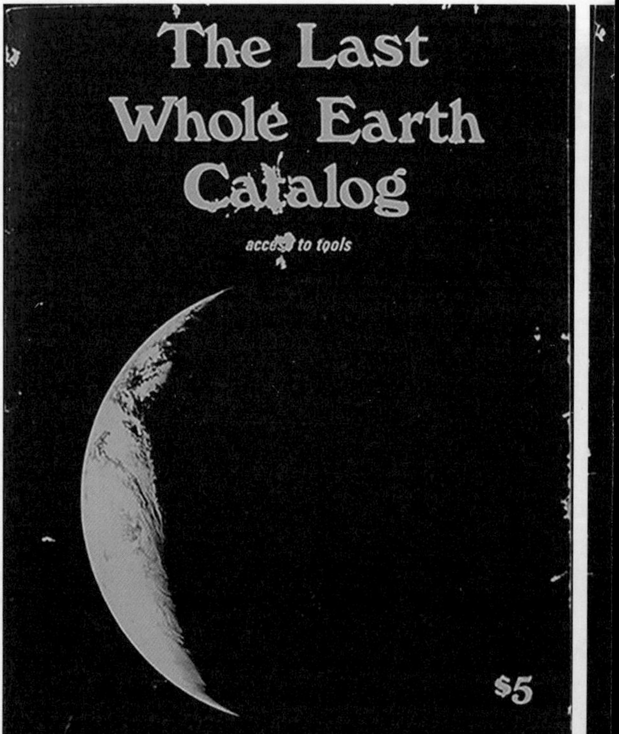

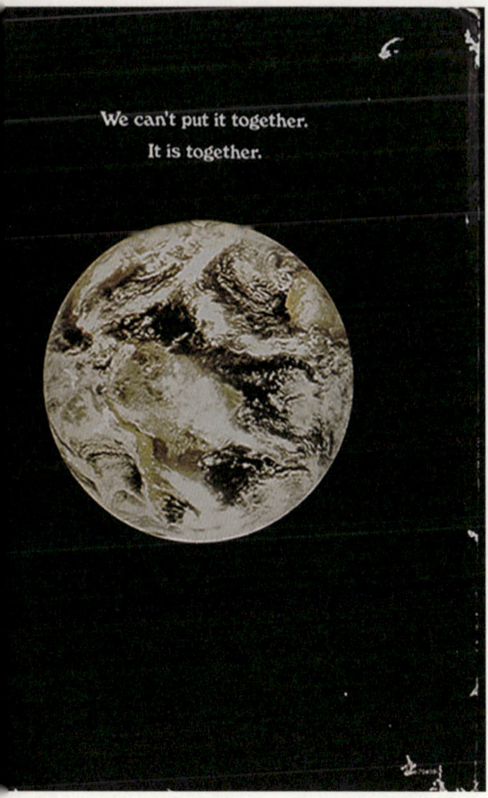
We can't put it together.
It is together.

For the back cover of the *Last Supplement of the Whole Earth Catalog* from 1971, editors Paul Krassner and Ken Kesey presented a cutaway globe against a pink background, showing part of it ripped apart by earthworms and undergrowth breaking through the surface. In this edition, *Whole Earth Catalog* founder Stewart Brand claimed that the magazine's iconic cover images of Earth reinforced objectives of the countercultural movement that had been gaining momentum since the late 1960s.[24] (courtesy Stewart Brand and the Department of Special Collections and University Archives, Stanford University Libraries)

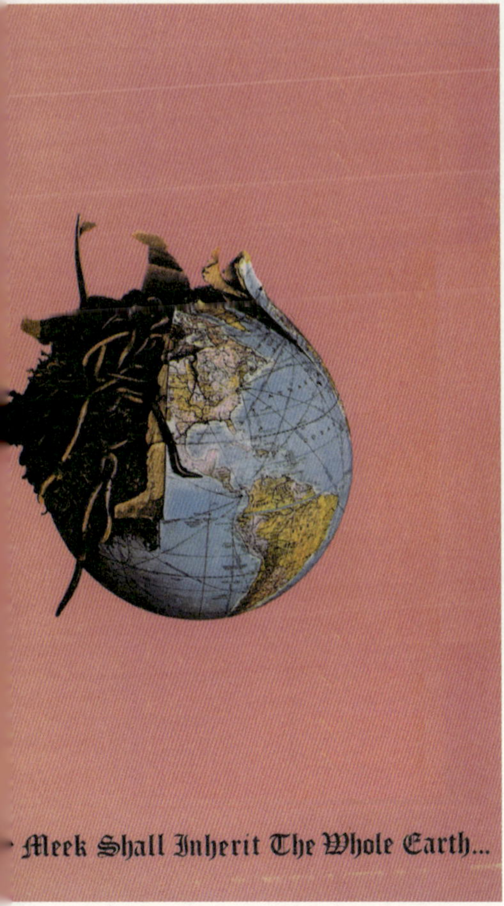
Meek Shall Inherit The Whole Earth...

Fascinated by the Moon landing in 1969 as well as the images disseminated by NASA of Earth seen from space, architect Alessandro Poli from the avantgarde architecture practice Superstudio developed the *Interplanetary Architecture Project* in 1971 for thinking about design in a more expanded domain to include not only terrestrial structures, but just as well those extending beyond into outer space; exhibited at the S-Space Mondial Festival in Florence in 1971 and published in *Casabella* in 1972. (courtesy Superstudio / Alessandro Poli)

SUPERSTUDIO

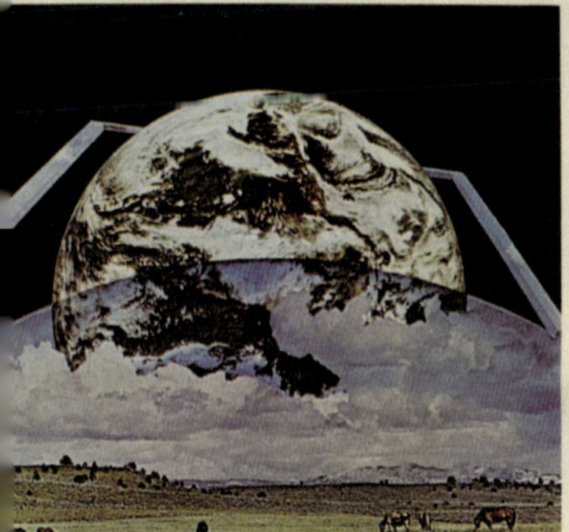

TTURA INTERPLANETARIA • 1971

The Blue Marble, reportedly one of the most reproduced photographs in history, was taken by the Apollo 17 crew on December 7, 1972; though the original image was oriented with the Geographic South Pole at the top, NASA released this seminal portrait of the Earth upside down to restore the familiar view with North on top. (courtesy National Aeronautics and Space Administration)

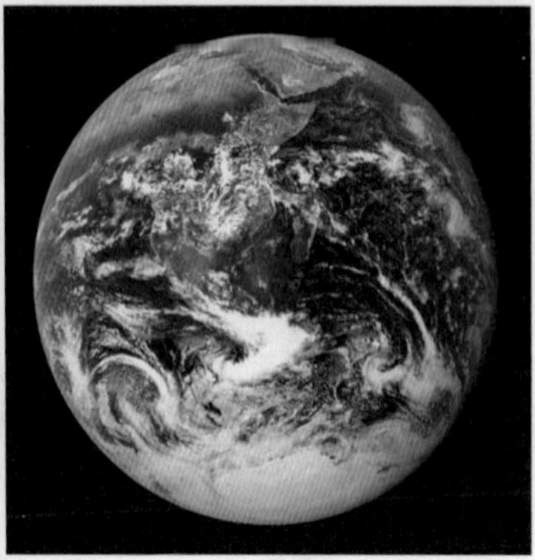

Cover of the first draft edition of *The Limits to Growth* prior to its official publication in 1972, which, using exponential indices, set out to predict the number of years until the world would run out of various resources and to reflect on the impact this depletion would have on human activity; MIT research project team: Donella H. Meadows, Dennis L. Meadows, Jørgen Randers, William W. Behrens III, et al.[25] (courtesy Club of Rome and Universe Books, New York)

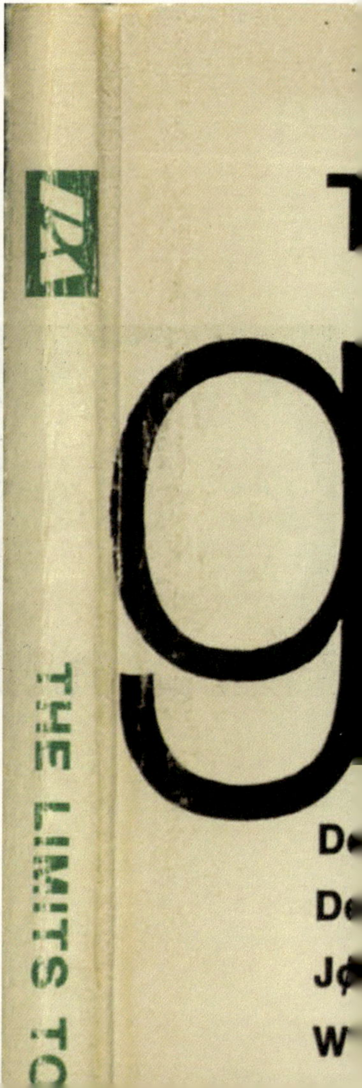

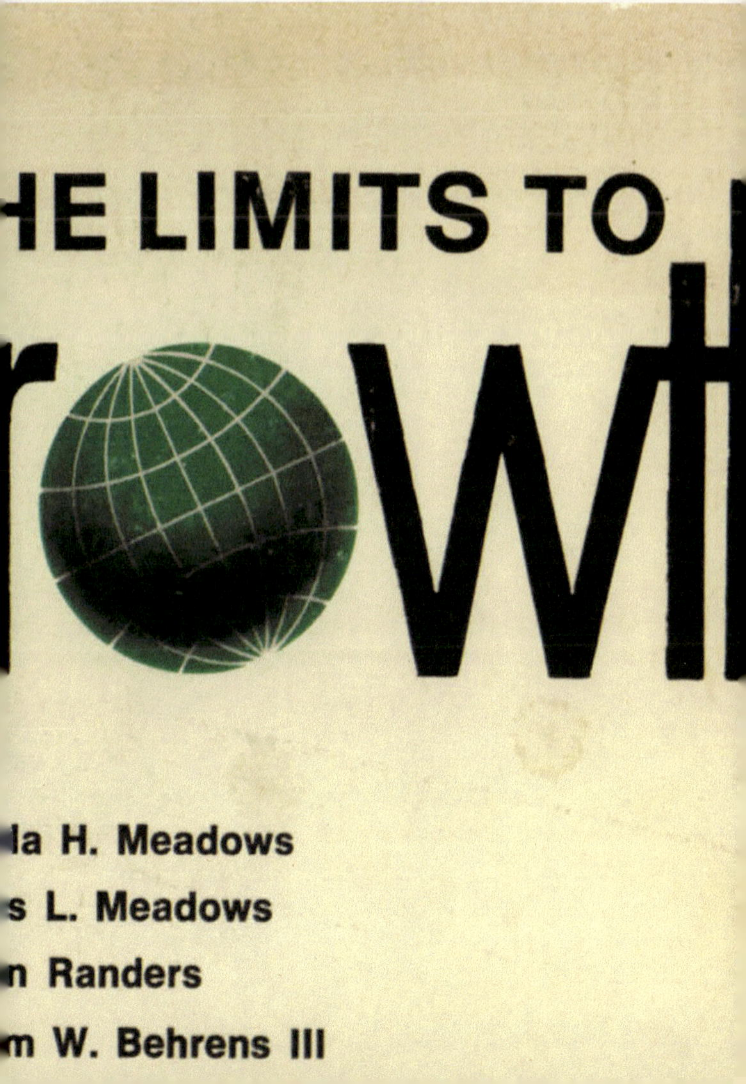
HE LIMITS TO
la H. Meadows
s L. Meadows
n Randers
m W. Behrens III

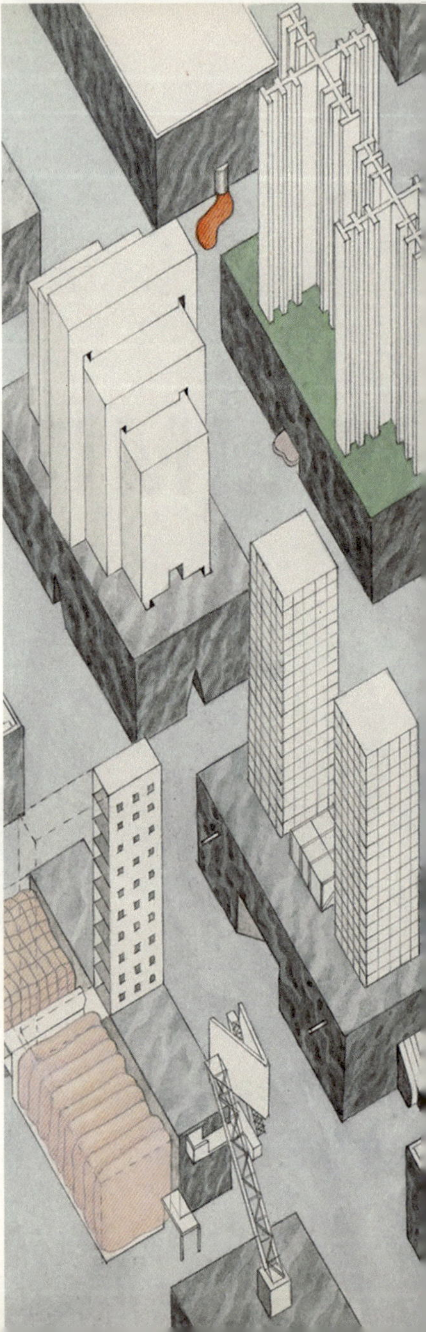

Drawing by Rem Koolhaas and Madelon Vriesendorp with Zoe Zenghelis produced as part of *The City of the Captive Globe Project* that captures the tension between the relentless regularity of the Manhattan grid and the delirious freedom of formal expression afforded to its architecture, 1972. (courtesy Scale Picture Library and Museum of Modern Art, New York)

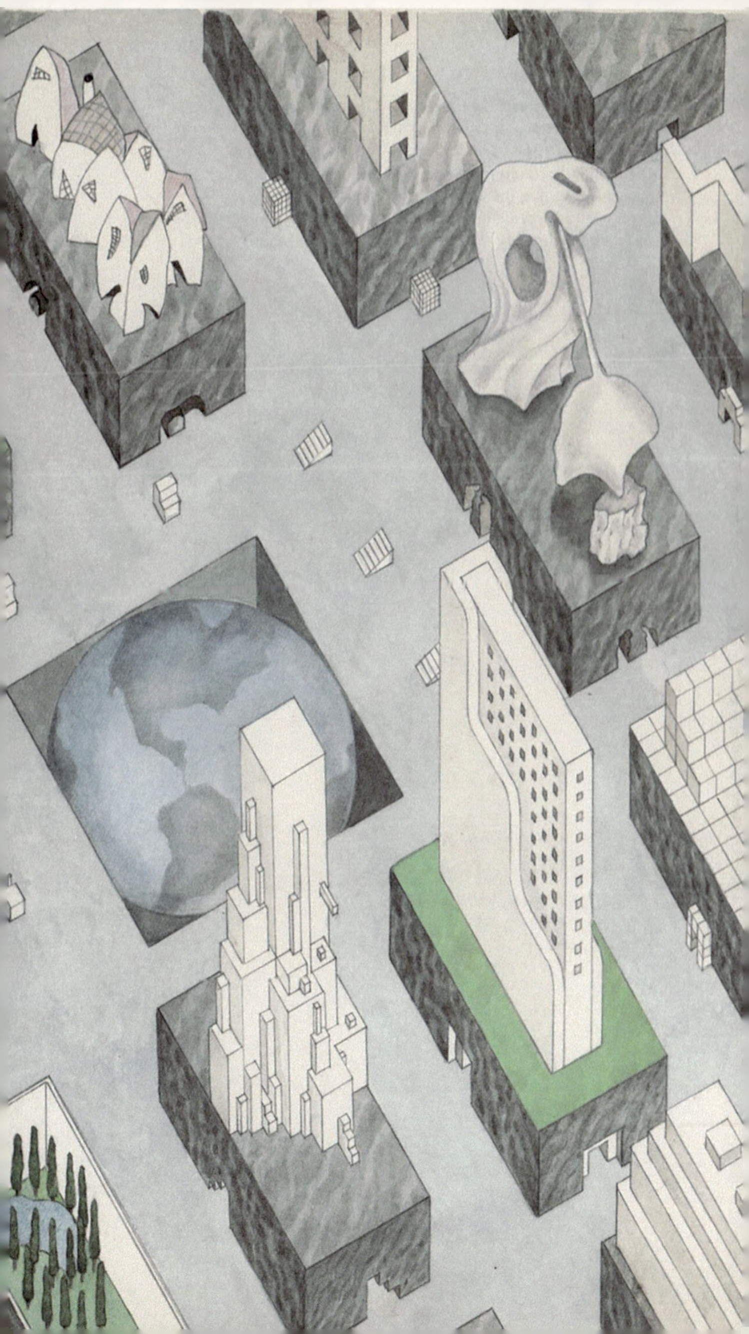

Chiasa Madre by architect Ludovico Quaroni is a church in the Sicilian town of Gibellina that was completely rebuilt after a 1968 earthquake; the church is comprised of, among other Platonic volumes, a large spherical space surrounding the altar and featured in a short film by Petra Noordkamp entitled *La madre, il figlio e l'architetto* from 2012.[26] (photograph Giona Andreani)

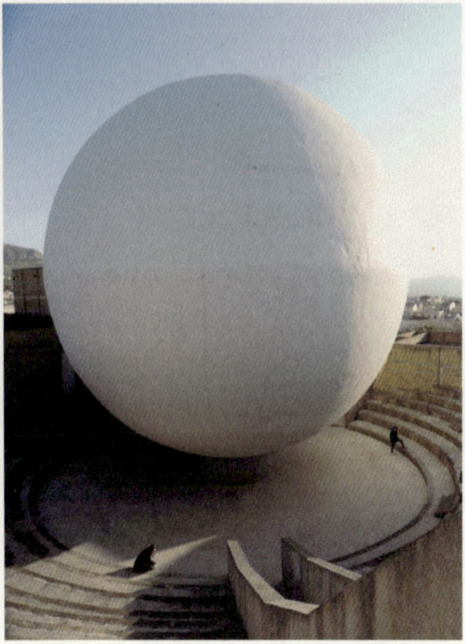

A few years after having named the computer the "Machine of the Year," *Time Magazine* named Earth "Planet of the Year" in 1989 to address mounting concerns about environmental deterioration; *Time Magazine* commissioned Javacheff Christo and his partner Jeanne-Claude to produce an image for the cover depicting the world's precarious situation, an image inspired by the earlier work *Wrapped Globe* from 1968 that showed a shrink-wrapped globe bound by ropes.[27] (courtesy Time Magazine)

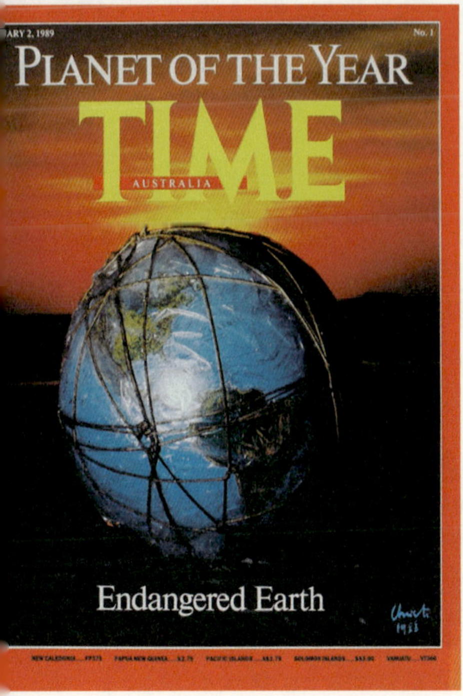
ARY 2, 1989
No. 1
PLANET OF THE YEAR
TIME
AUSTRALIA
Endangered Earth

Cover of *The New Yorker* from April 1990 celebrating the 20th anniversary of Earth Day by depicting a polluted planet on the shoulders of Atlas wearing a gas mask; by that time, Earth Day had become a worldwide event, with some 140 countries and 200 million people demonstrating for the cause of a healthier environment, an event that paved the way for the 1992 United Nations World Summit in Rio de Janeiro. (illustration Bob Mankoff, courtesy The New Yorker)

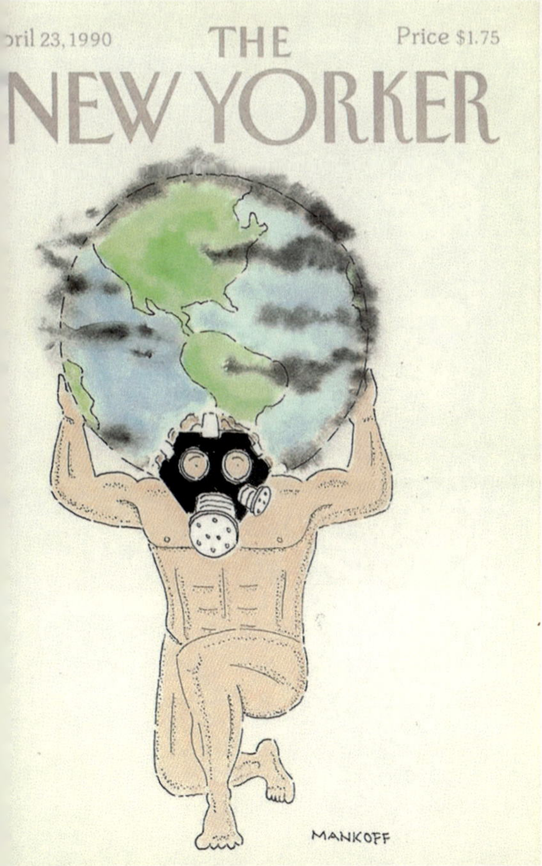
pril 23, 1990
THE
Price $1.75
NEW YORKER
MANKOFF

Eurosphäre auf Erde by Guido Nussbaum (ca. 1997) depicting a dismembered Europe scattered across the globe. (photograph Schweizerisches Institut für Kunstwissenschaft, SIK-ISEA Zurich, courtesy Alice Jaeckel and Guido Nussbaum)

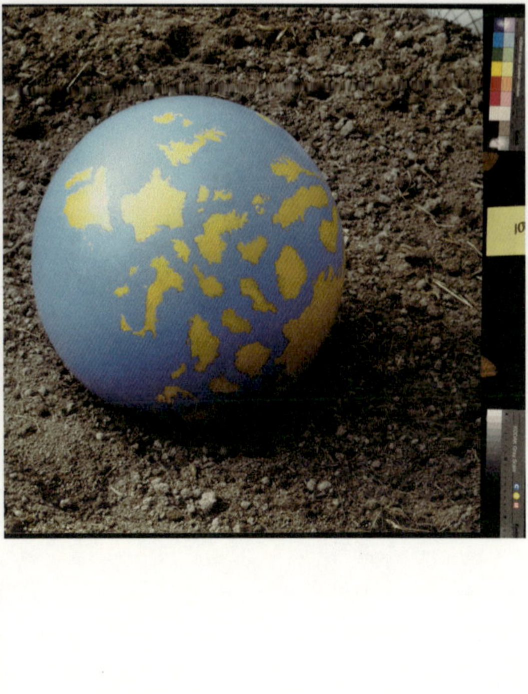

Mark Lombardi, drawing entitled *World Finance Corporation and Associates, ca. 1970-84: Miami, Ajman, Bogota, Caracas (7th version)* providing a graphic narrative of international power networks and financial transactions, 1999. (courtesy Scale Picture Library and Mark Lombardi Networks)

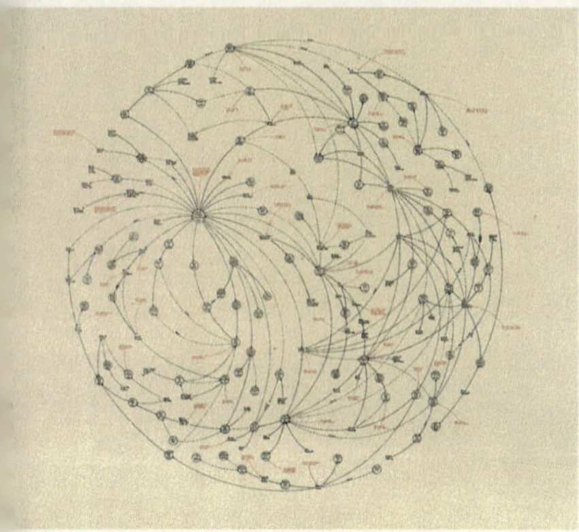

Photograph by Andy Sacks of migrant worker assembling a globe on the production line in a factory in Broadview, Illinois, 2001; the image was subsequently used for the cover of Kimberly Ann Elliott and Richard B. Freeman's 2003 book *Can Labor Standards Improve Under Globalization?*[28] (photograph Andy Sacks, courtesy Getty Images)

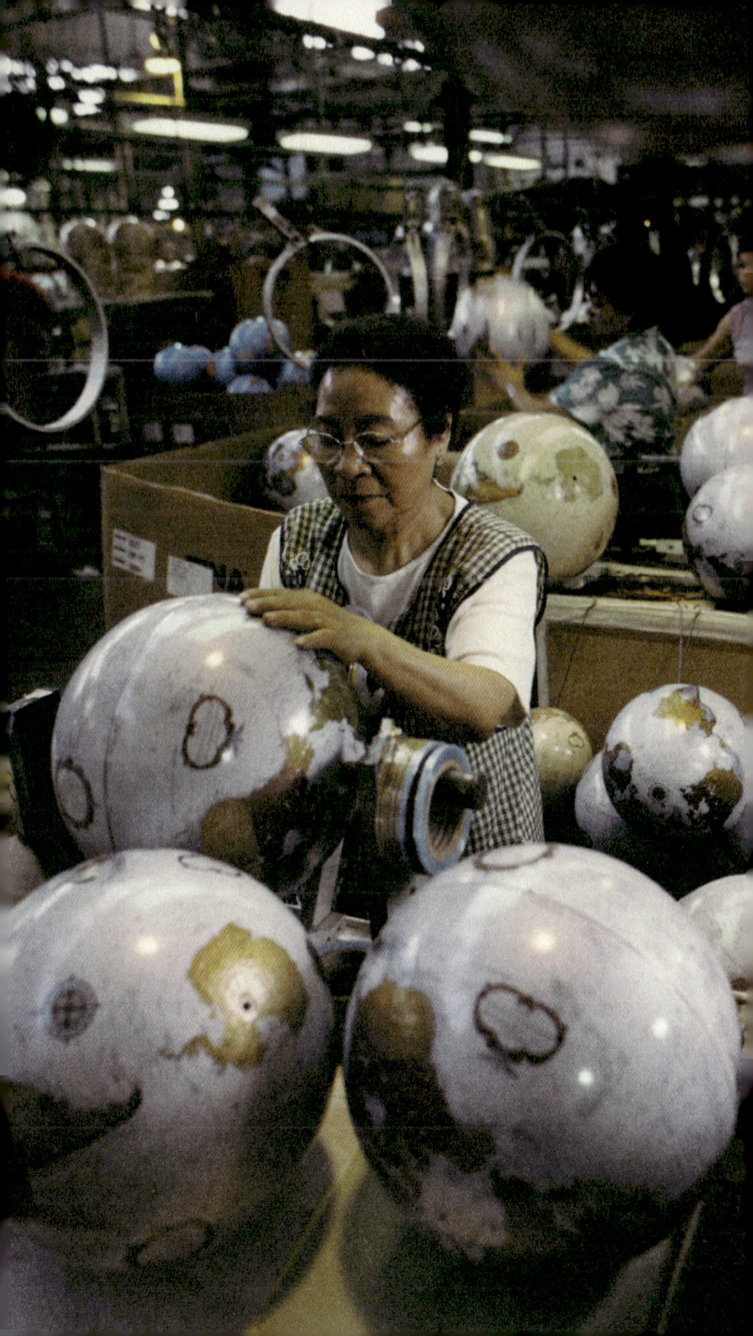

Virtual world on the homepage of Google Earth from 2005, by now the most popular geospatial application used for viewing the planet as a whole and in all its details; first developed by Keyhole Inc. in 2001, the search engine was purchased by Google in 2004 and released in a rebranded version in 2005; the company now dominates the global online market with approximately 70% of total shares, commanding a true monopoly on the digital perception of the world, except in those places where Google-based search engines are blocked.[29] (courtesy Google Inc.)

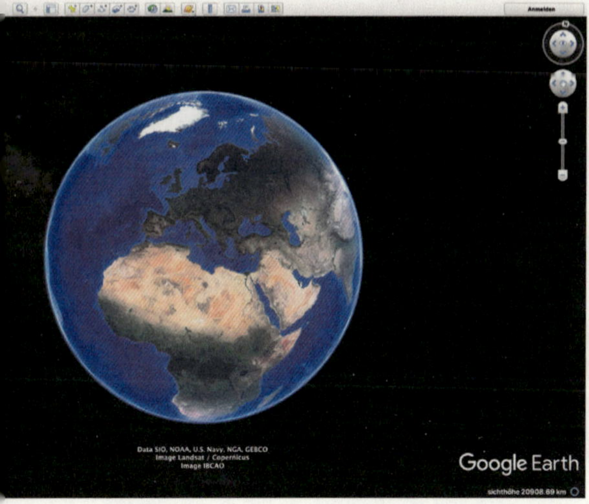
Anmelden
Data SIO, NOAA, U.S. Navy, NGA, GEBCO
Image Landsat / Copernicus
Image IBCAO
Google Earth
sichthöhe 20908.69 km

Image of the Earth geoid from the Gravity Recovery and Climate Experiment mission (GRACE) implemented by the National Aeronautics and Space Administration (NASA) and the German Aerospace Center (DLR) to precisely map Earth's ever-changing gravitational field, fluctuations mostly due to changes in how water is constantly being redistributed around the planet; the mission, lasting from 2002 to 2017, provided insights into interactions of the Earth's hydrosphere, atmosphere, and lithosphere, 2009. (courtesy NASA and Goddard Space Flight Center Scientific Visualization Studio)

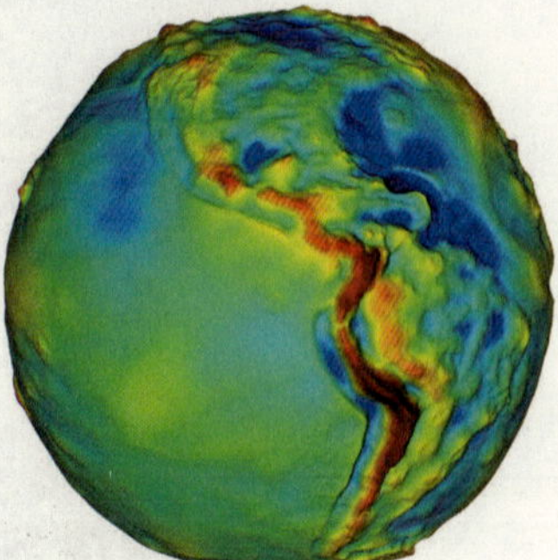

Outgrowth, by artist Thomas Hirschhorn, is a wall installation presenting 131 globes 'contaminated' by outgrowths made of brown adhesive tape; each globe is accompanied by a photograph of a particularly violent incident that occured in a specific place somewhere on the globe; all items are awkwardly lined up on storage shelves, 2005. (photograph Georges Meguerditchian, courtesy Centre Pompidou, MNAM-CCI /Dist. RMN-GP and Adagp, Paris)

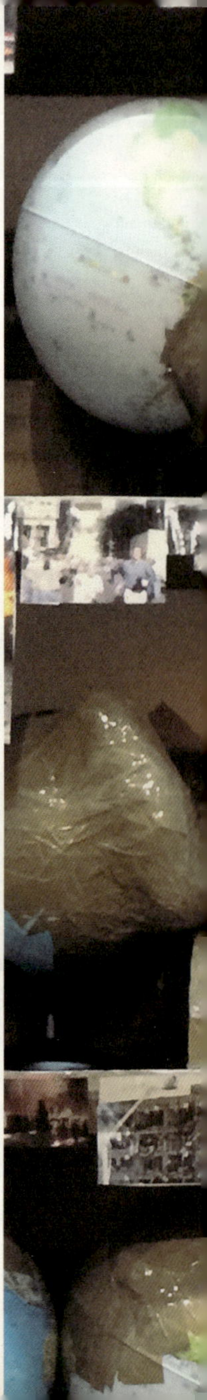

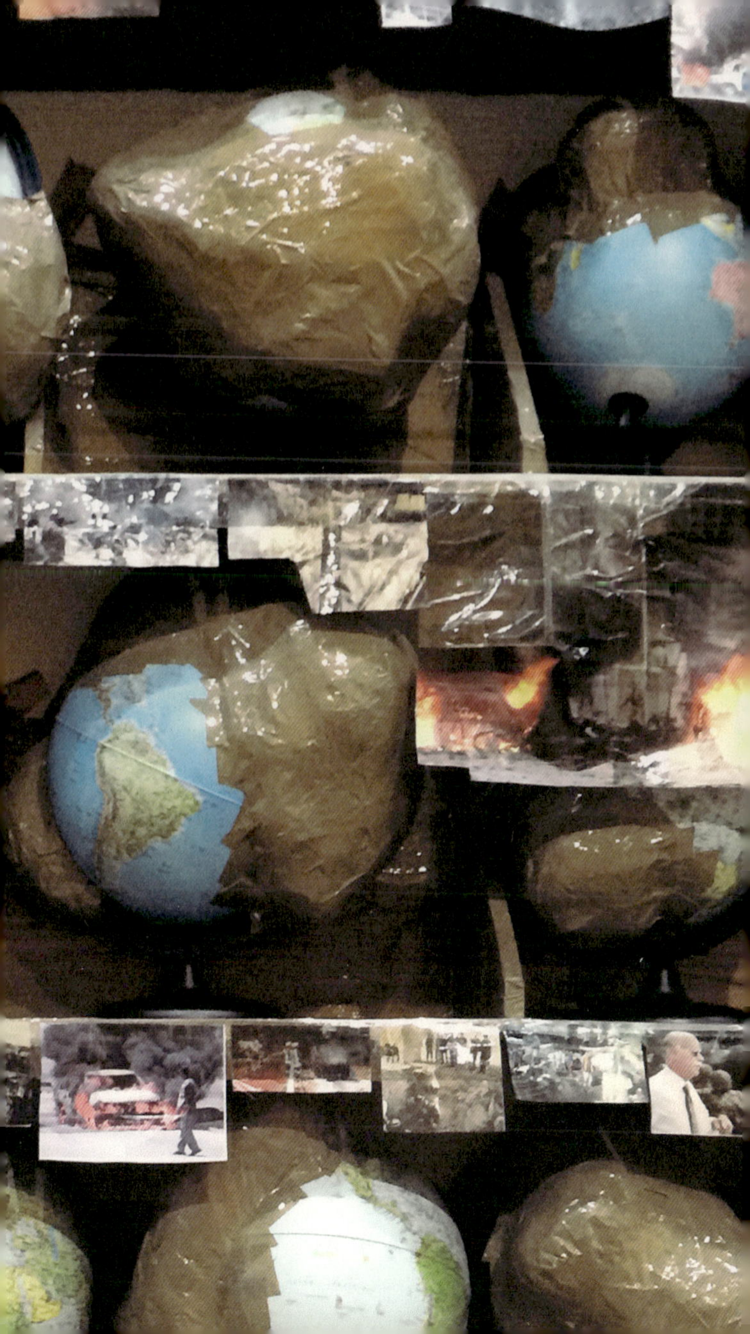

Data visualization by artists Scott Hessel and Gabriel Dunne from the 2005 UCLA project *Celestial Mechanics*, which, in taking its name from the branch of astronomy that studies the motion of objects in outer space, thematizes the urbanization of the sky by tracking the paths and functions of human-made aerial technologies, in this case weather balloons, GPS networks (from the United States and Russia), and low-Earth orbit satellites. (courtesy Scott Hessel and Gabriel Dunne)

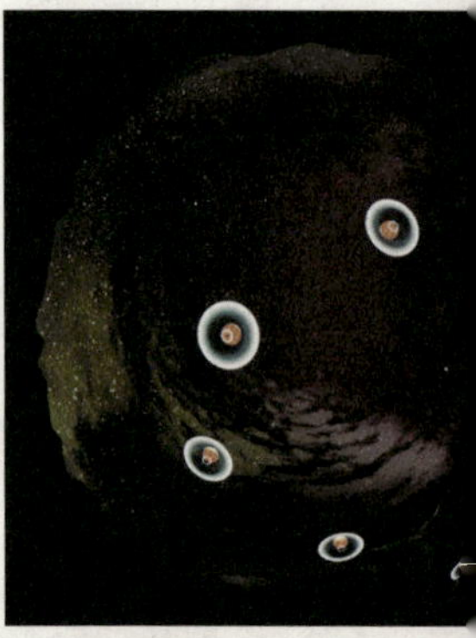

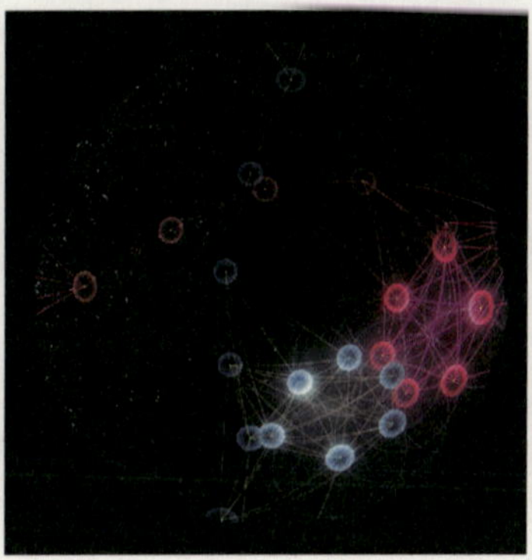

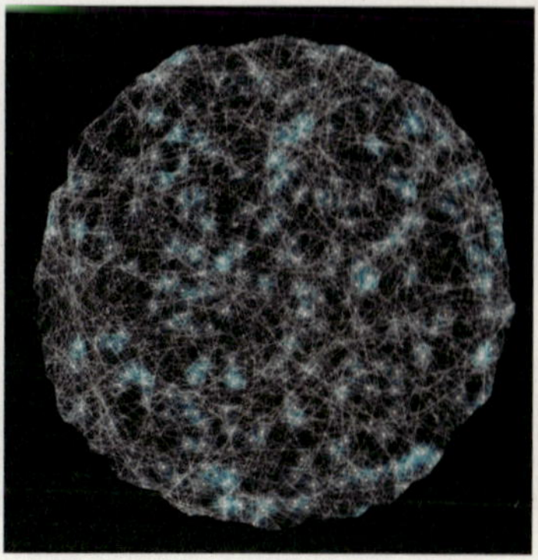

Data visualization by artist Aaron Koblin entitled *Flight Patterns* from 2005 showing air traffic over North America during a 24-hour period using information from the Federal Aviation Administration (FAA); the work was originally produced as part of the *Celestial Mechanics* project at UCLA with colleagues Scott Hessel and Gabriel Dunne that thematizes technologies of the sky. (courtesy Aaron Koblin)

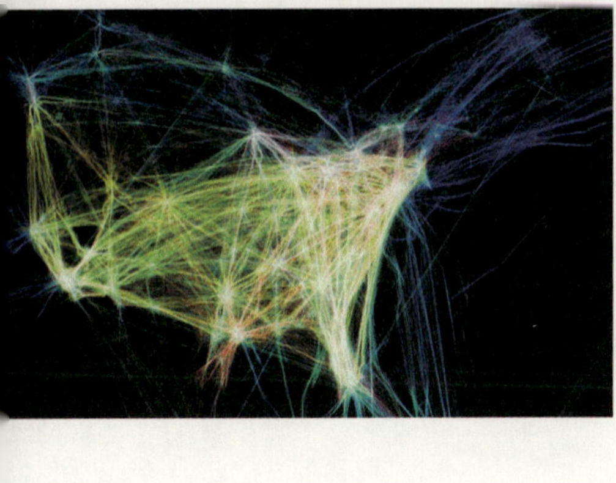

Map by Philippe Rekacewicz for *Le Monde diplomatique* entitled *la grande roue* showing a cartographic interpretation of the "plundering machinery" at work on the African continent under international development cooperation programs, 2007.[30] (courtesy Philippe Rekacewicz, visionscarto.net)

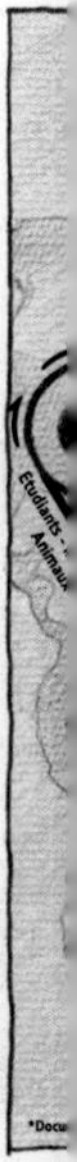

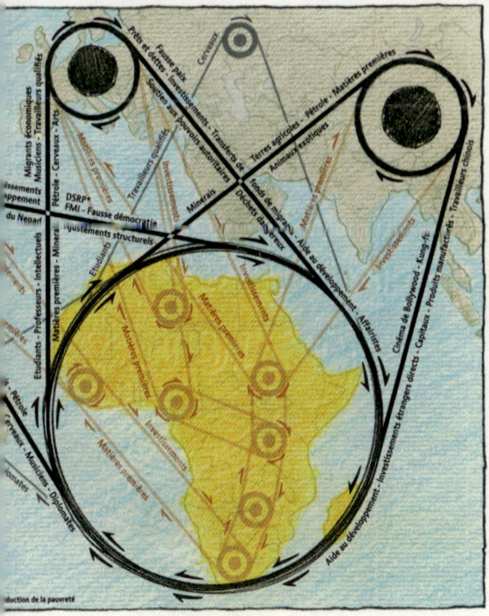
Prêts et dettes - Investissements - Transferts de fonds de migrants - Aide au développement - Affairistes
Fausse paix
Soutien aux pouvoirs autoritaires
Cerveaux
Terres agricoles - Pétrole - Matières premières
Animaux exotiques
Déchets dangereux
Migrants économiques
Musiciens - Travailleurs qualifiés
Pétrole - Cerveaux - Arts
Matières premières
Travailleurs qualifiés
Investissements
Minerais
DSRP*
FMI - Fausse démocratie
du Nepad
Ajustements structurels
Etudiants
Etudiants - Professeurs - Intellectuels
Matières premières - Minerai
Pétrole
Cerveaux - Musiciens - Diplomates
Matières premières
Investissements
Matières premières
Investissements
Cinéma de Bollywood - Kung-fu
Capitaux - Produits manufacturés - Travailleurs chinois
Aide au développement - Investissements étrangers directs

Multimedia installation *EXIT* by Diller Scofidio + Renfro at the Palais de Tokyo in Paris (commissioned by the Fondation Cartier) showing geo-coded data on demographic shifts, forced migration, natural disasters, rising sea levels, and increasing deforestation rates – a show which intentionally coincided with the COP21 UN conference in Paris in late 2015; the presentation was based on earlier reflections by Paul Virilio in a 2008 interview on the state of Earth in the age of climate change (courtesy Fondation Cartier pour l'art contemporain)

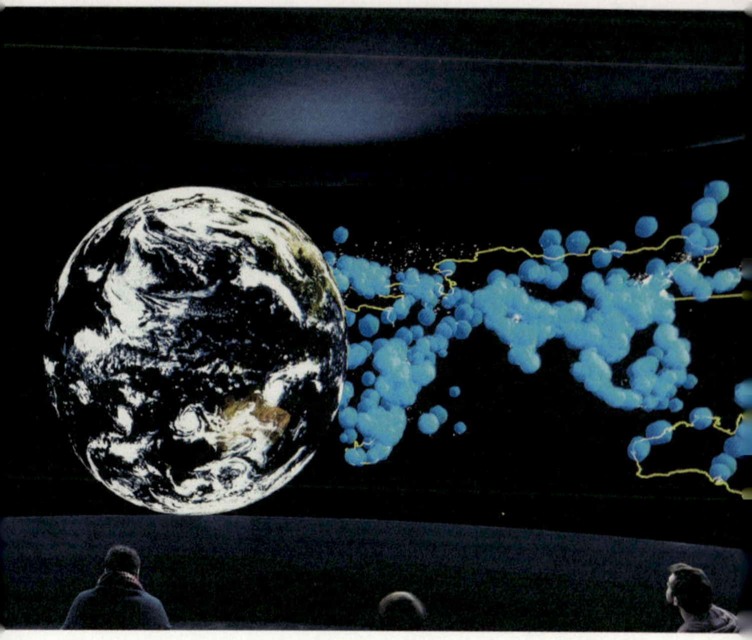

Model of the Earth geoid from 2011 based on a state-of-the-art gradiometer mapping made by the Gravity Field and Ocean Circulation Explorer (GOCE) sent into orbit by the European Space Agency (ESA) in 2009 to study variations in the Earth's gravitational field as well as deep-water current circulation in the oceans; the colors in the image represent deviations in the intensity of gravity that range from +100m to -100m relative to an ideal sphere; the mission, which provided at the time the best representation of global gravity, ended in 2013. (courtesy ESA/HPF/DLR)

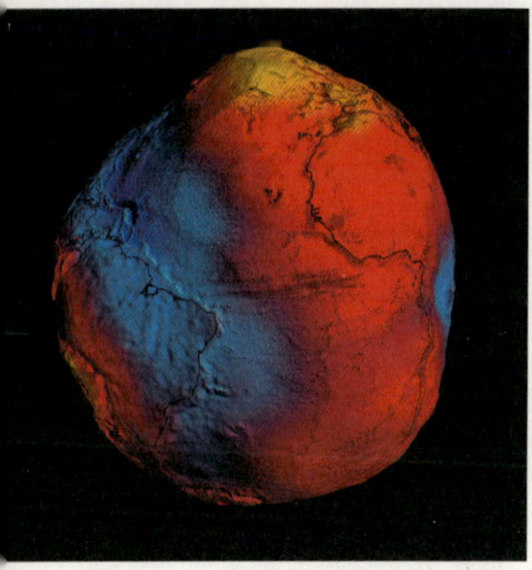

Stadtigel (urban porcupine), CNC model of a spherical city that is endless and without beginning, with neither periphery nor center – a model reportedly missing since 2010; produced by the research collaborative KAISERSROT from ETH Zurich for the 4th International Architecture Biennale Rotterdam, 2009-10. (Markus Braach, Benjamin Dillenburger, Oliver Fritz, Ludger Hovestadt, Alexander Lehnerer, Tom Pawlofsky, Mathias Bernhard, et al., Department of Architecture, ETH Zurich)

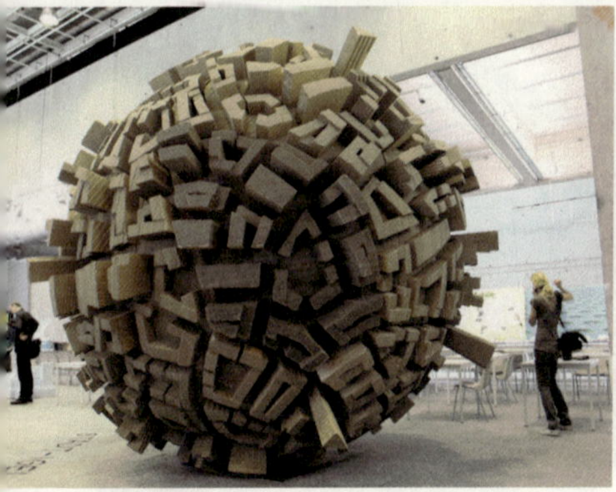

Series of *Endlessly Open Cities* that accompanied the installation *Stadtigel* at the 4th International Architecture Biennale Rotterdam, 2009-10: *City of the Continuous Roof*, Renee Ciolino; *City of the Eternal Commute*, Andrew Brosseit; *City of the Golden Globe*, Ryan Hollon; *City of the Seamless Runabout*, Janis Rucins.[31] (Alexander Lehnerer, ETH Zurich, in collaboration with the University of Illinois at Chicago)

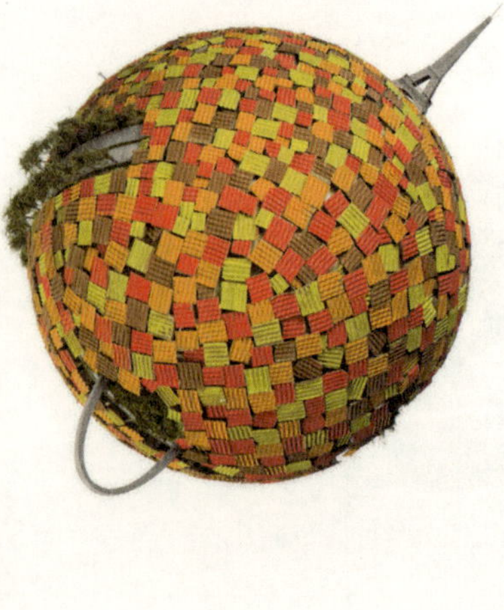

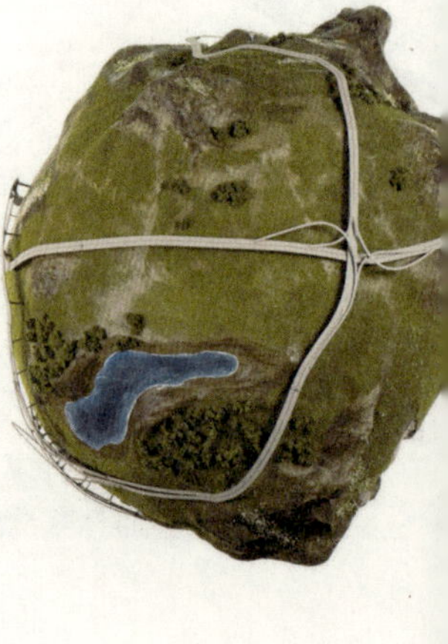

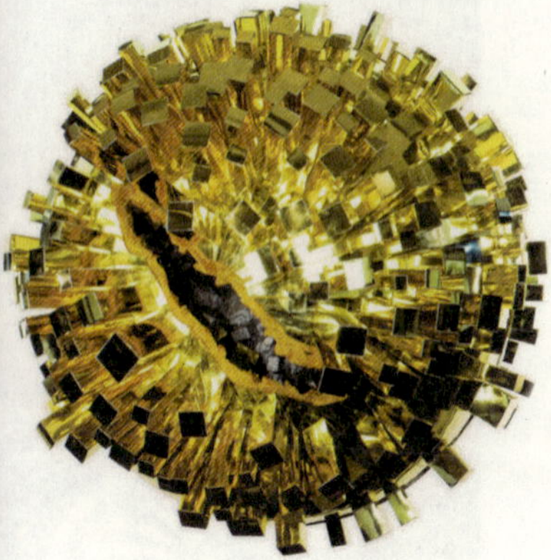

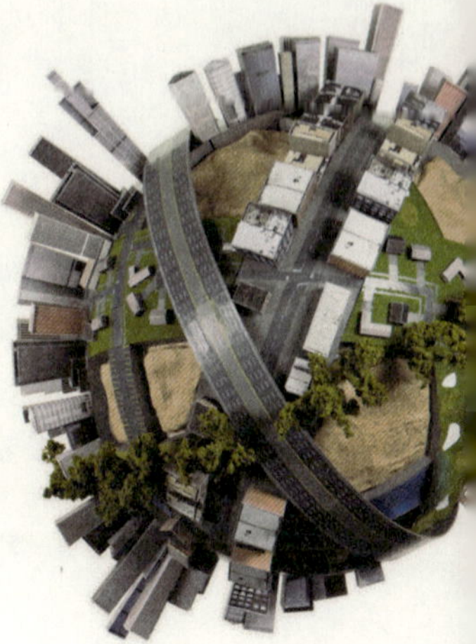

Satellite photograph from 2010 of *The World* just off the shore of Dubai showing an archipelago of islands arranged in the shape of the world as seen on a map; the venture was developed by Nakheel Properties that at the time was one of the most spectacular real estate holdings of Dubai World.[32] (courtesy National Aeronautics and Space Administration)

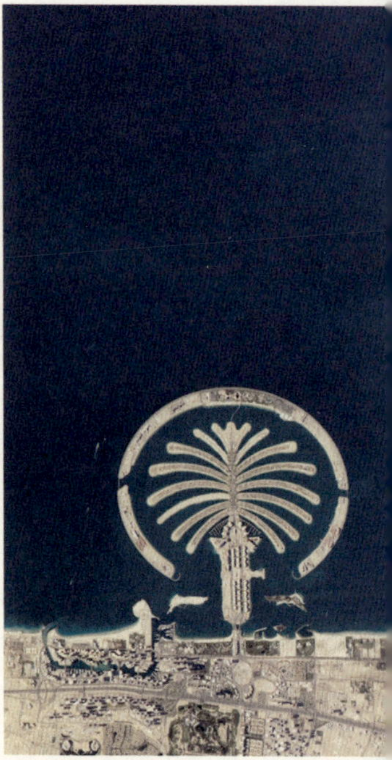

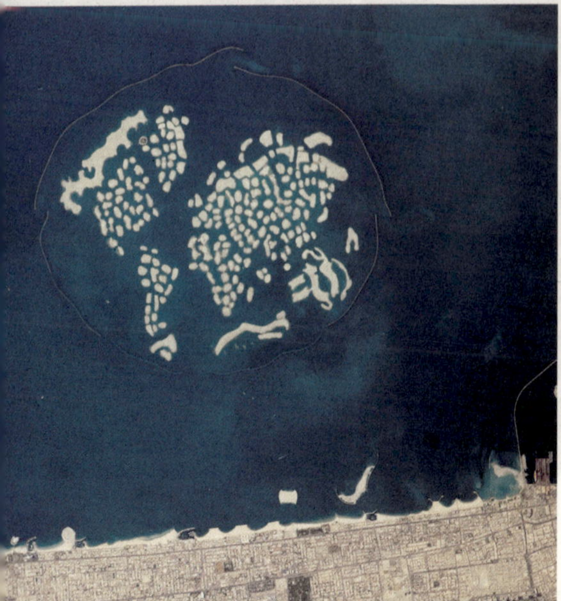

Video stills from the 5-minute visual narrative *'Urban' World*? on varying urban epistemologies, including the "Ecumenopolis" concept by Constantinos Doxiadis, the "Global City" network proposed by Saskia Sassen among others, and the model of "Planetary Urbanization" put forth by Neil Brenner and Christian Schmid, 2016 (courtesy Neil Brenner, Daniel Ibanez, Nikos Katsikis, et al., Urban Theory Lab, Graduate School of Design, Harvard University)

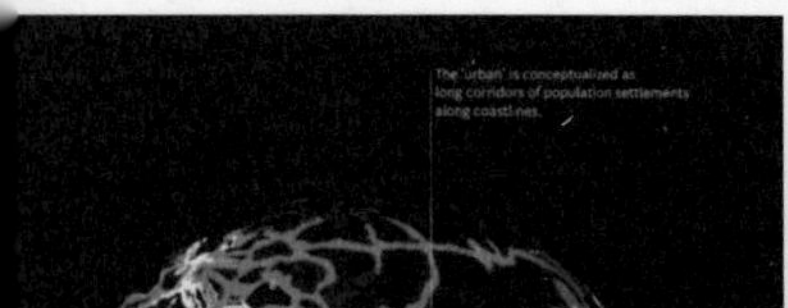

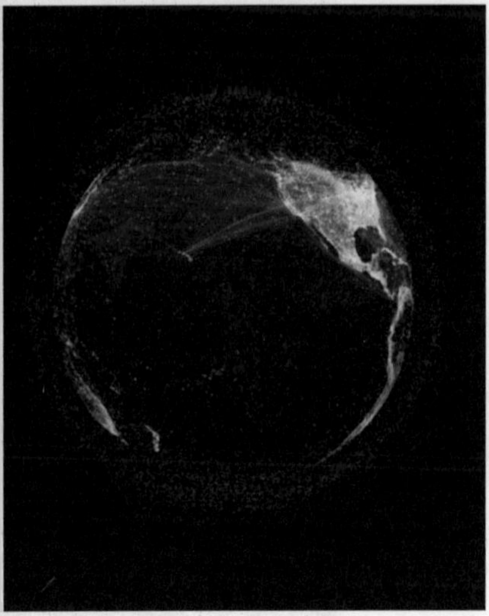

Rendering of a prospective renewable global energy grid prepared by OMA/AMO for the 2011 *Energy Report* in collaboration with WWF and Ecofys, proposing that the world's energy needs could be met entirely by renewable sources by 2050.[33] (courtesy OMA)

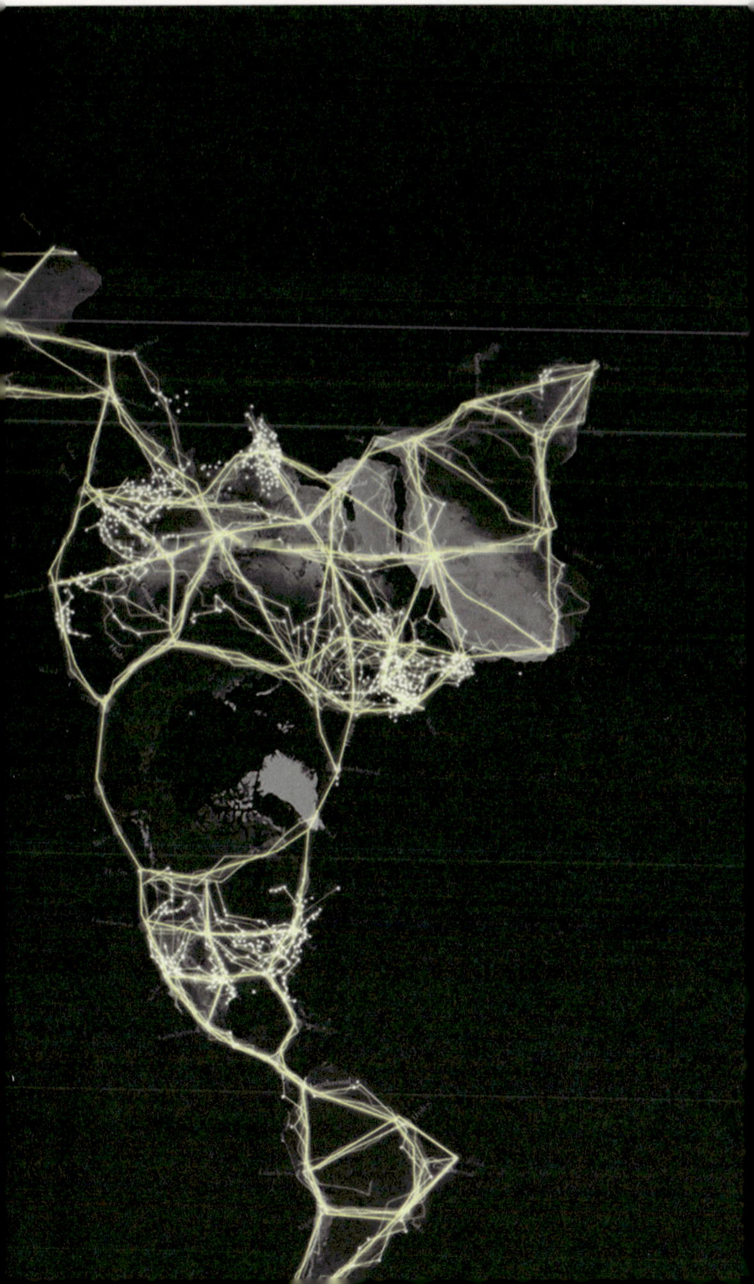

Cover image made for an issue of *The Economist* entitled “Welcome to the Anthropocene: Geology’s New Age” showing planet Earth as a technical contraption somewhere in the process of being put together and taken apart, with its substructure partially exposed and jet streams escaping from its interior, May 28, 2011.[34] (illustration by Jon Berkeley for The Economist)

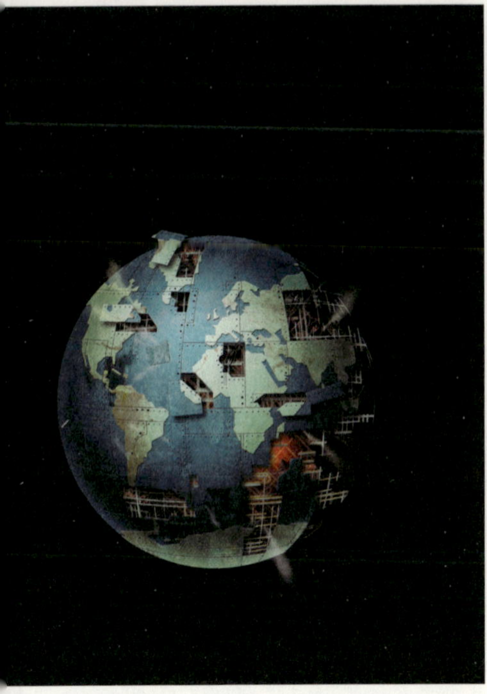

OmniGlobe at ETH Zurich, a spherical screen on which different projections visualize dynamic processes such as plate tectonics, climate change, terrestrial magnetism, or even the surface of other planetary bodies; in essence, the model constitutes a generic globe that can be animated by data sets to present Earth's geophysical properties in realistic simulations for teaching and research purposes, 2011-12.[35] (photograph Radek Brunecky, courtesy *focus*Terra, ETH Library, and Institute of Cartography and Geoinformation, ETH Zurich)

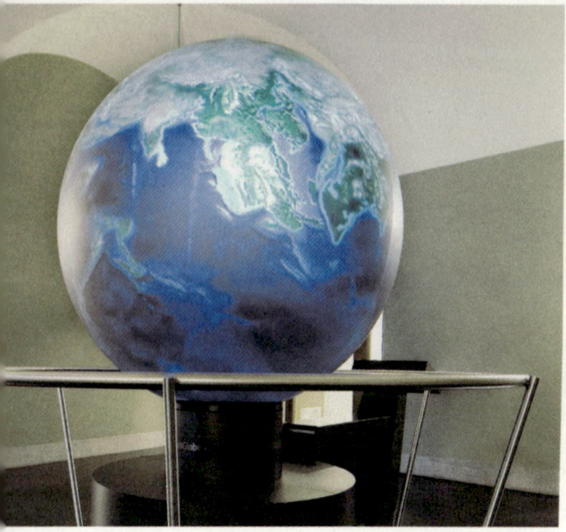

Photograph from 1969 by artist-architect François Dallegret taken of his wife Judith late in her pregnancy with twins, the swollen belly exposed as the maiden human bubble within a vacuum of the darkened background; the image seems to anticipate Peter Sloterdijk's assertion some 30 years later in the first volume of his *Spheres* trilogy entitled *Bubbles* that women's bodies are prenatal 'apartments' – the ur-dwelling of our species; the image was part of a 2012 exhibition shown at the Institute for the History and Theory of Architecture (gta), Department of Architecture, ETH Zurich.[37] (courtesy François Dallegret, Alessandra Ponte, Laurent Stalder, and Thomas Weaver)

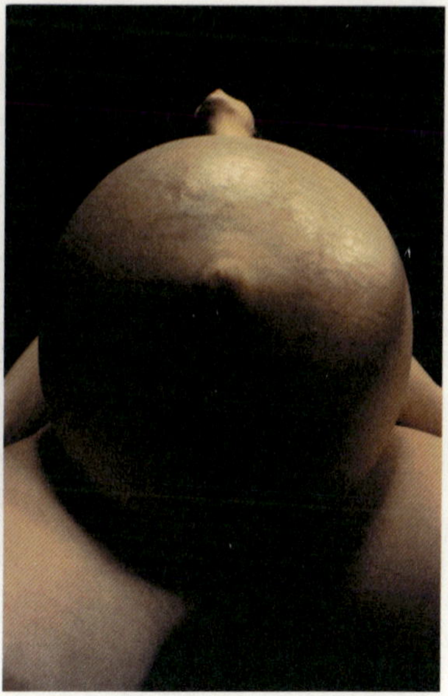

Temporary inflatable structure that hovers over an outdoor concert venue and appears as an enigmatic object in the medieval setting of Saint-Prex on Lake Geneva, one that would ideally take flight to become a celestial body in its own right as reflected in its name, *luna*, 2012 version.[36] (photograph Joël Tettamanti, courtesy Dieter Dietz, Atelier de la conception de l'espace or ALICE, EPFL)

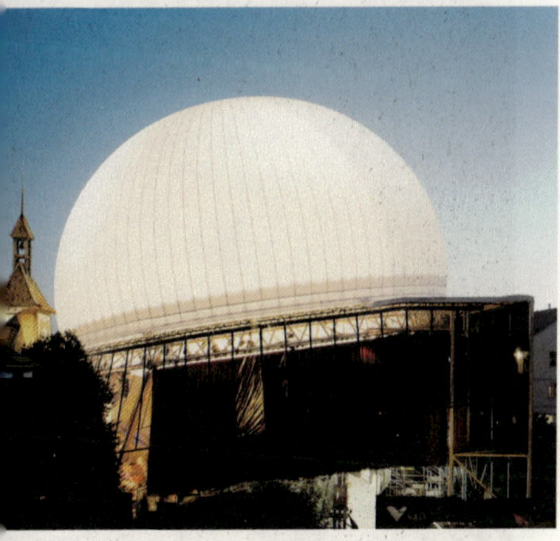

Animated digital climate model of the Earth depicting Representative Concentration Pathways (RCPs) or time-dependent projections of atmospheric greenhouse gas accumulation and its effect on global warming; in this case the climate future scenarios are for a +1°C-world and +4°C-world by 2100 (blue curve and red curve, respectively), Institute for Atmospheric and Climate Science, Department of Environmental Systems Science, ETH Zurich, 2012. (courtesy Reto Knutti, Oliver Stebler, Jan Sedlácek, Urs Beyerle)

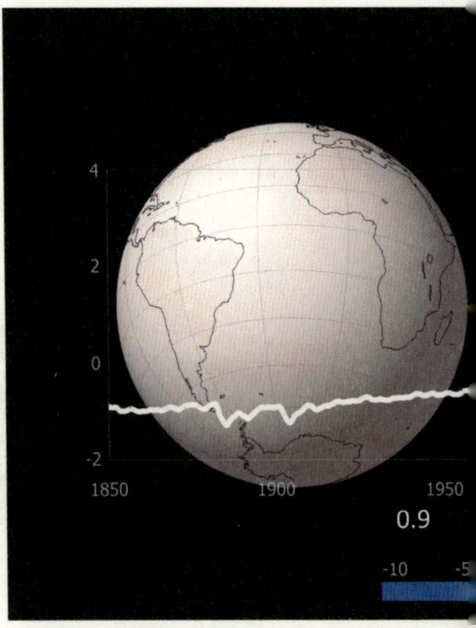

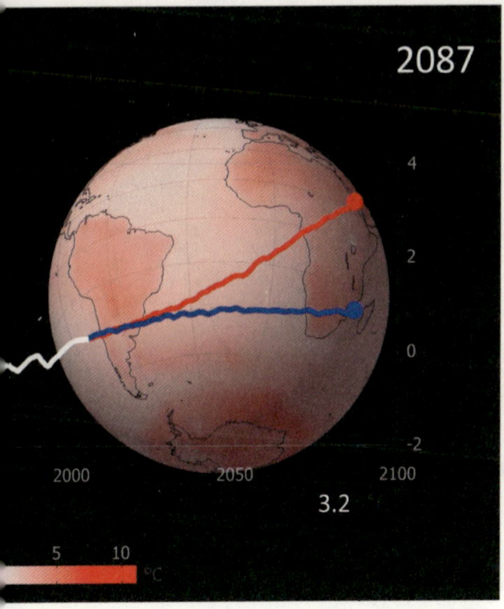
2087
4
2
0
-2
2000
2050
2100
3.2
5
10
°C

Images of the so-called *Black Marble*, or Earth at night, made from a composite of photographs taken by the Suomi National Polar-orbiting Partnership satellite (Suomi NPP) jointly managed by the National Aeronautics and Space Adminitration (NASA) as well as the National Oceanic and Atmospheric Administration (NOAA); the resulting images, showing the extent of global urbanization patterns as of 2016 (revised from 2012 data), were mapped onto updated *Blue Marble* images of Earth to provide a seemingly realistic, albeit fabricated, view of the planet. (courtesy NASA and NOAA)

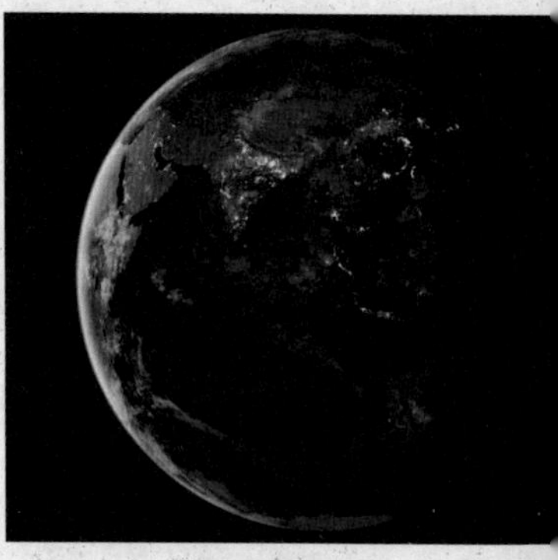

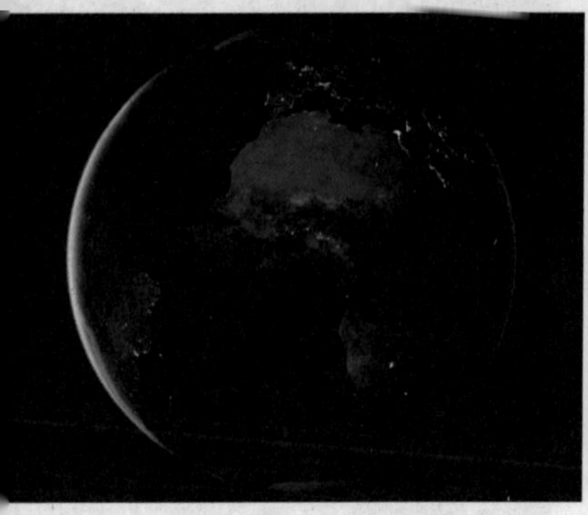

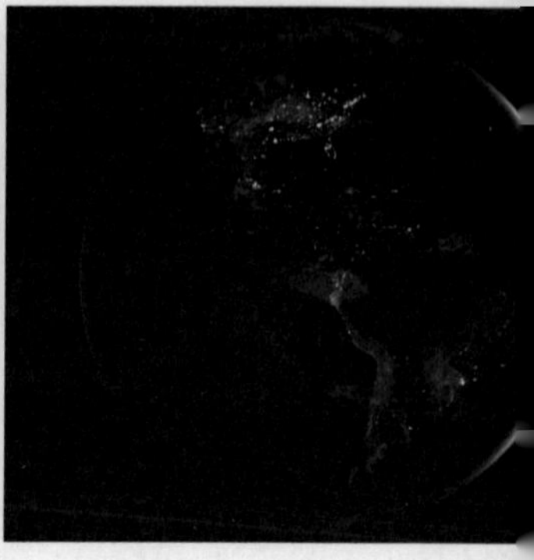

Computer-generated artwork *space debris I* from 2012 by artist Michael Najjar depicting the so-called “orbital cascade” of space junk that, having accumulated since the mid-1960s, will most likely increase in the coming decades and expand the ‘urbanization’ of Earth’s outer atmosphere; the artwork was created in collaboration with the Institute of Aerospace Systems of the Technical University Braunschweig.[38] (courtesy Michael Najjar)

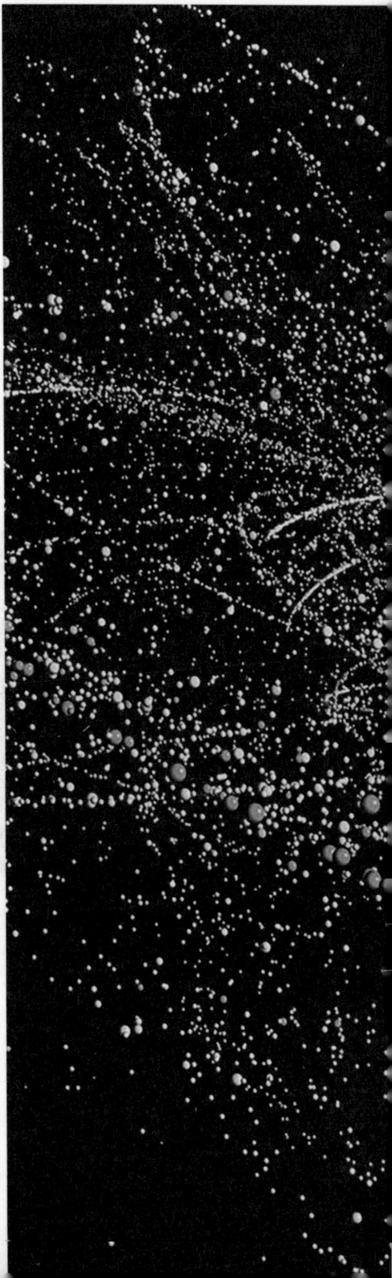

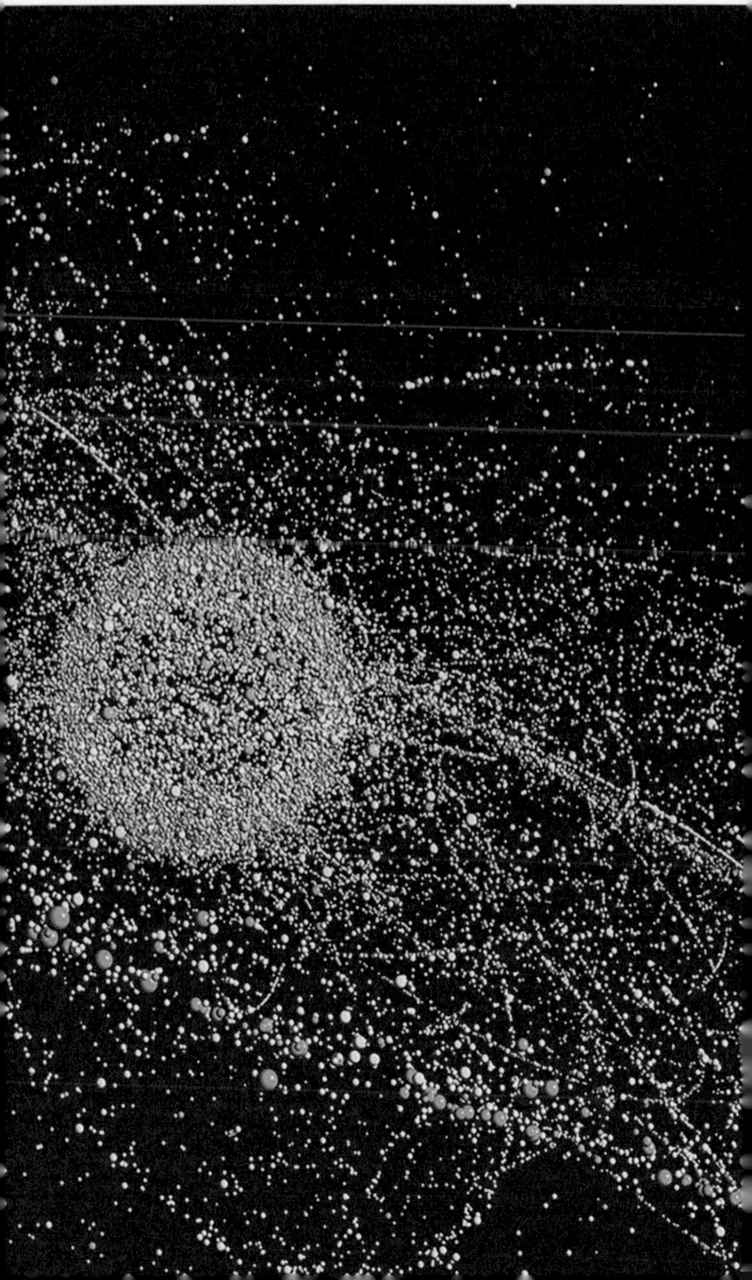

Digital simulation of the magnetic field in the Earth's core, which fluctuates and occasionally undergoes polarity reversals over a time span of 40,000 years; Institute for Geophysics, Department of Earth Sciences, ETH Zurich, and Swiss National Supercomputing Centre, 2012.[39] (courtesy Jon Rotvig, Andrew Jackson, and Jean Favre)

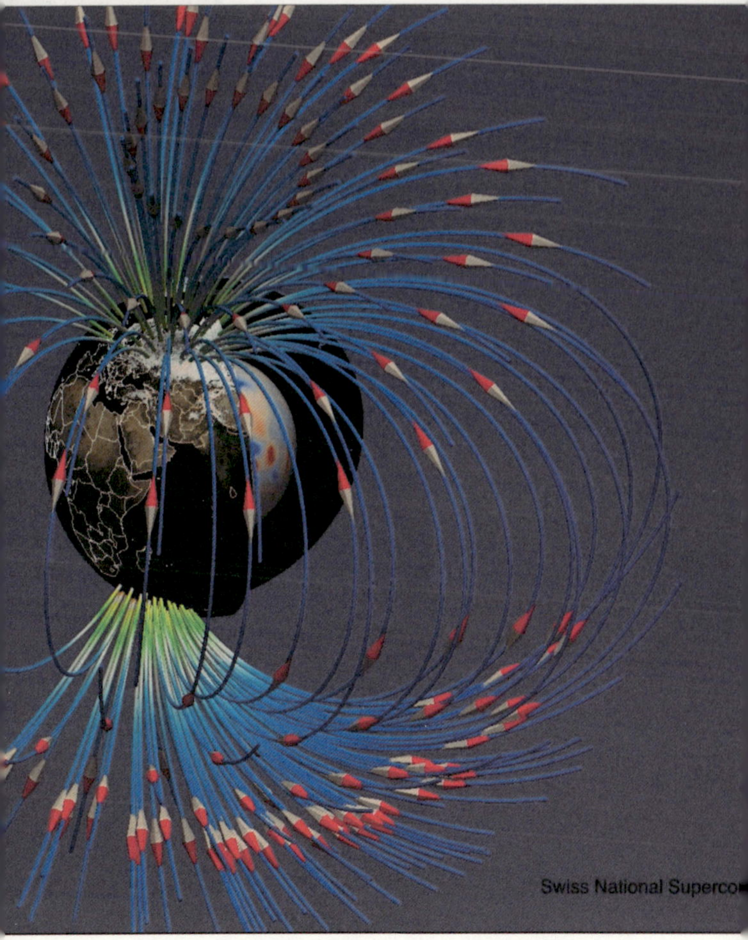
Swiss National Superco

Spherical micro-tissue of a cell derived from primary rat neurons, cultured, stained, and imaged in a microfluidic channel to disclose the structure of the smallest unit of life on the globe; finalist of the 2013 MicroTAS Art in Science Awards; Olivier Frey et al., Bio Engineering Laboratory, Department of Biosystems Science and Engineering, ETH Zurich, 2013. (courtesy Olivier Frey, Bio Engineering Laboratory)

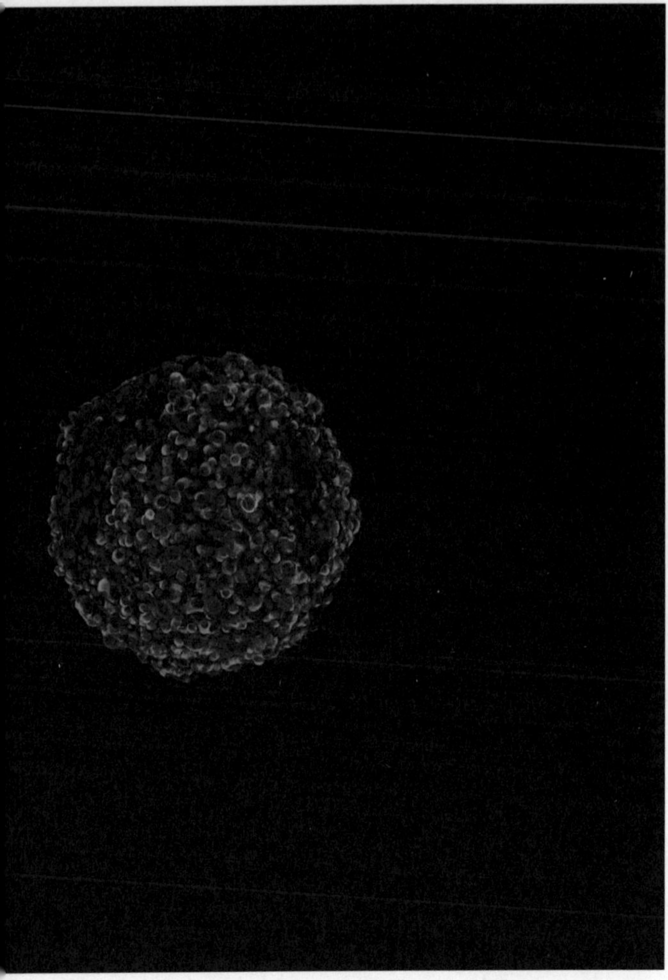

Still from a video accompanying the installation *Terra Forming: Engineering the Sublime* by Adam Lowe and Jerry Brotton for the 2014-15 exhibition *Antropocène Monument* organized by Bruno Latour at Les Abattoires in Toulouse; the model displayed at the exhibition showed a waterless world which was gradually flooded, as if to highlight the fear of rising ocean levels brought on by anthropogenic transformations of the globe.[40] (courtesy Adam Lowe and Jerry Brotton; Factum Arte, Madrid, London, and Milan)

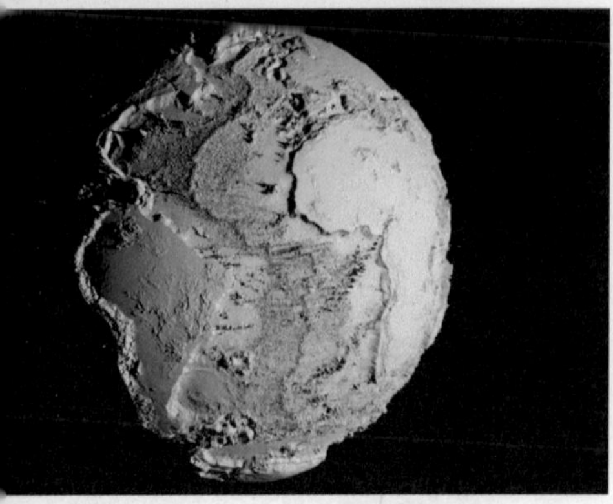

Photograph of Karin Sander's *Gebrauchs-Bild 169b* installed at the exhibition space Atelier Amden overlooking the Walensee in Switzerland, 2014-2019; the series of use- or patina-paintings registers processes of the surrounding environment as they work upon the canvas. (photograph Studio Karin Sander, courtesy Atelier Amden / Pro Litteris)

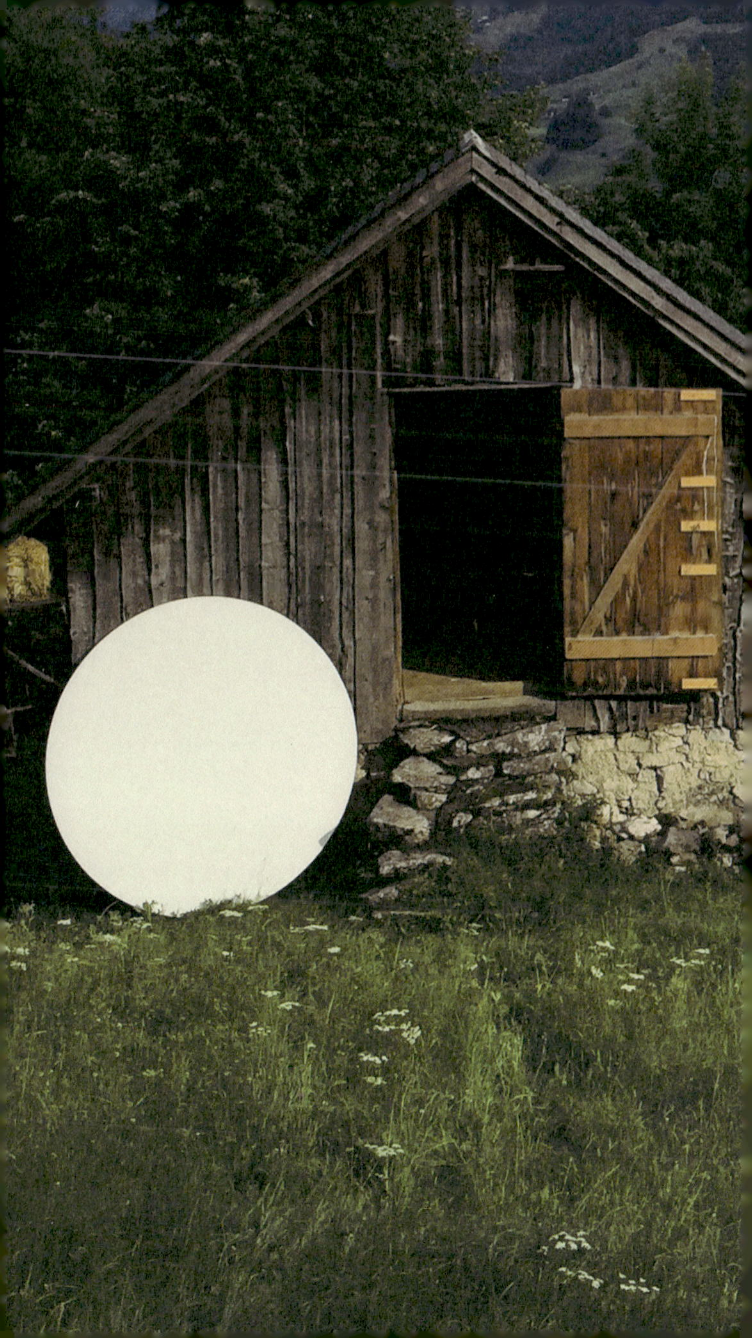

Screenshots from a computer simulation of plate tectonics and mantle convection showing Earth's geophysical fluid dynamics, with zones where plates from the terrestrial surface are pulled into the mantle (blue/green) as well as the upwelling of hot rock material from the deep mantle (red), Institute of Geophysics, Department of Earth Sciences, ETH Zurich, 2015-19.[41] (courtesy Paul Tackley et al.)

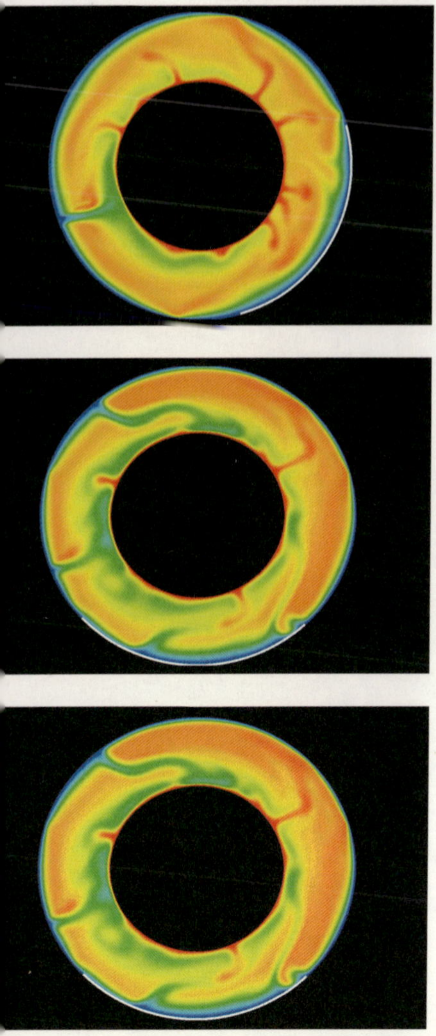

Visualization by Alexandra Arènes entitled *Anthropogenic Earth* from 2015 rendering the planet in terms of the human impact on the world environment, showing soil depletion, wasteland zones, melting ice caps, excavated topographies, plantation distribution, abandoned infrastructures, urbanization patterns, toxic clouds, and so forth; produced for Bruno Latour's 2016-17 performance lectures *Inside*. (courtesy Alexandra Arènes, Atelier SOC)

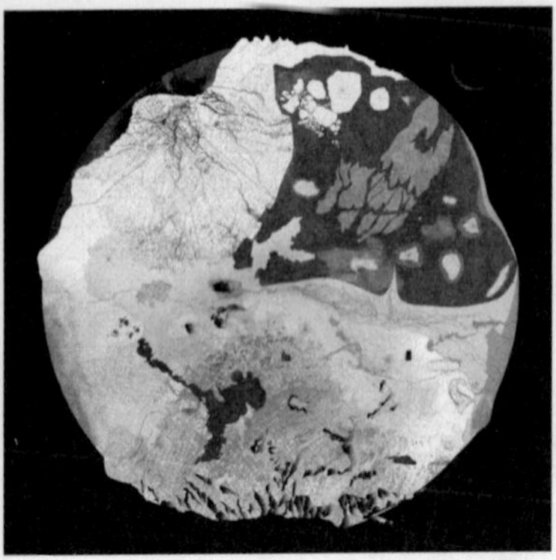

Animated visualization by Axelle Grégoire entitled *Recomposition* from 2016 showing a patchwork-globe of recomposed fragments of cities, freeways, mines, airplanes, runways, shipping lanes, industrial zones, patches of dead sea, toxic clouds, and so forth meant to draw attention to the disjunctive orders of the planet; produced for Bruno Latour's 2016-17 performance lectures *Inside*. (courtesy Axelle Grégoire and Sonia Levy)

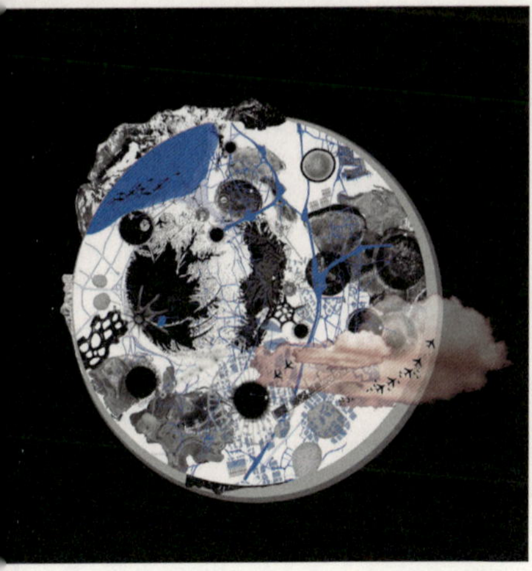

Animated notation by Alexandra Arènes entitled *Cartogenesis of the Territory of Belval* from 2016 made of a site in the Ardennes region of France showing the multiple traces of human and non-human actors as they interact to produce a distinct territorial assemblage; produced for Bruno Latour's 2016-17 performance lectures *Inside*. (courtesy Alexandra Arènes, Atelier SOC)

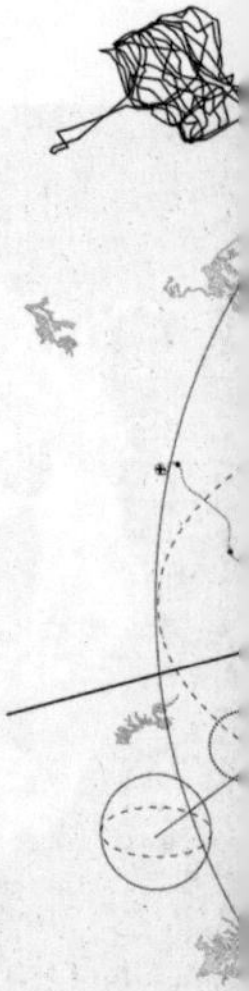

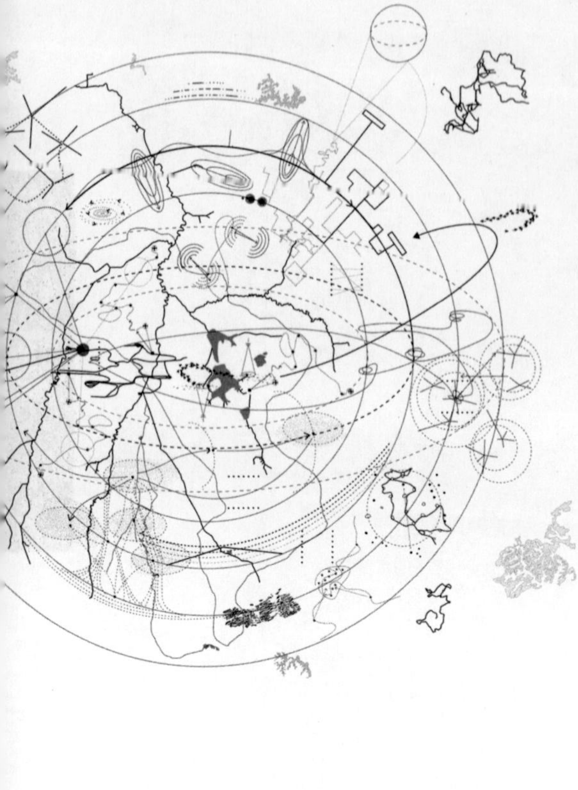

Still from an animation of Earth used by foreign policy reseacher Parag Khanna in a 2016 TED Talks presentation addressing what he calls in his book *Connectography* the "global connectivity revolution;" according to his contentious argument, the world has been transformed from one based on the "political geography" of sovereign nations to one operating according to the borderless "functional geography" of monetary circuits and material flows.[42] (courtesy Parag Khanna and TED Talks)

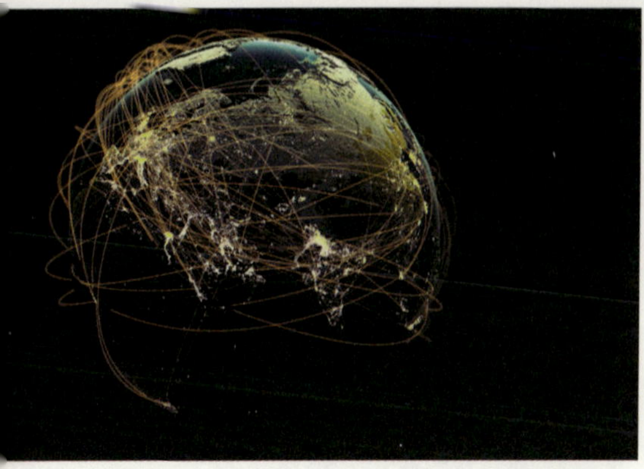

Screenshots from the *Swiss World Atlas Interactive*, a web-based digital complement to the analogue version of the 2017 *Swiss World Atlas* that allows viewers to access layers of information, including local temperature, concentrations of radioactivity, and levels of economic development for venues on Earth; the research was initiated by the Conference of the Directors of Education of the Swiss Cantons and conducted at the Institute of Cartography and Geoinformation of ETH Zurich, 2010-17; see www.schweizerweltatlas.ch.[43] (courtesy Lorenz Hurni, Stefan Räber, et al., Institute of Cartography and Geoinformation, Department of Civil, Environmental and Geomatics Engineering, ETH Zurich)

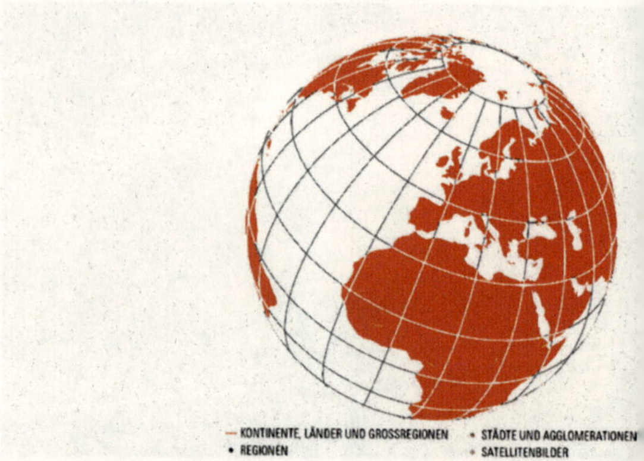

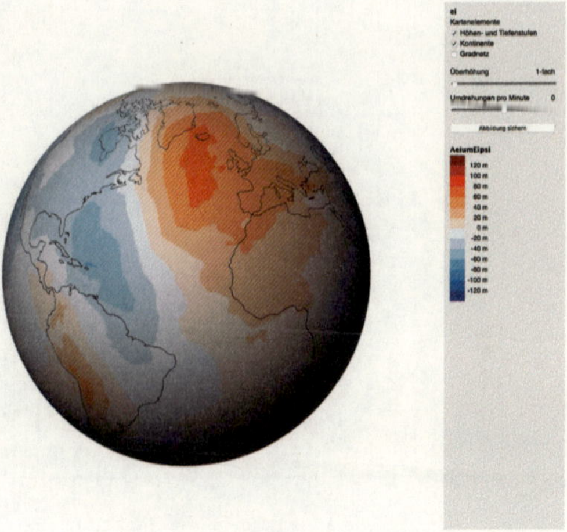
ei
Kartenelemente
✓ Höhen- und Tiefenstufen
✓ Kontinente
Gradnetz
Überhöhung
1-fach
Umdrehungen pro Minute
0
Abbildung sichern
AeiumEipsi
120 m
100 m
80 m
60 m
40 m
20 m
0 m
-20 m
-40 m
-60 m
-80 m
-100 m
-120 m

Montages by Rania Ghosn and El Hadi Jazairy from the architecture collective Design Earth made for an installation entitled "Blue Marble Circus" from 2017 showing the Blue Marble image of Earth – icon of the environmental movement – set within the scaffolding of the Pantheon – icon of architectural perfection; the blue shrink-wrapped globe within the Pantheon's spherical volume calls attention to the the plastic footprint of humanity.[44] (courtesy Design Earth, Rania Ghosn and El Hadi Jazairy)

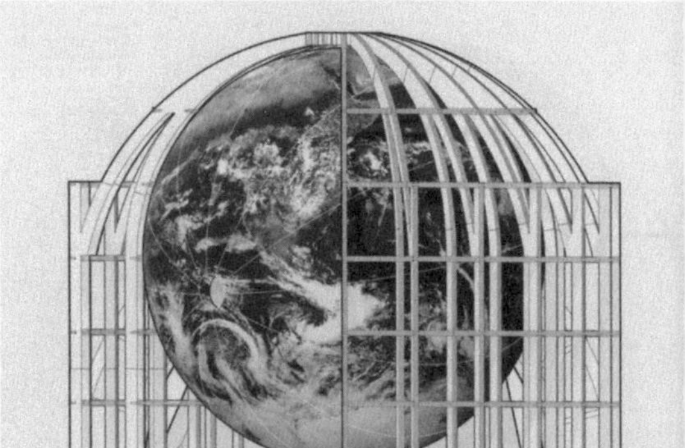

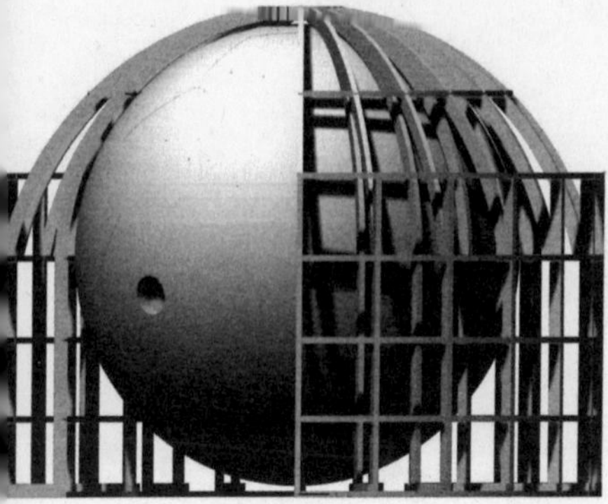

Drawings by architect Angelo Bucci from the University of São Paulo entitled *The Thin Layer* depicting the inhabitable realm of the Earth as a shallow zone of approximately 3.5 km in height; whereas the circumference appears as a single line, it is in reality comprised of two lines, the inner and outer circles of the thin layer. (courtesy Angelo Bucci)

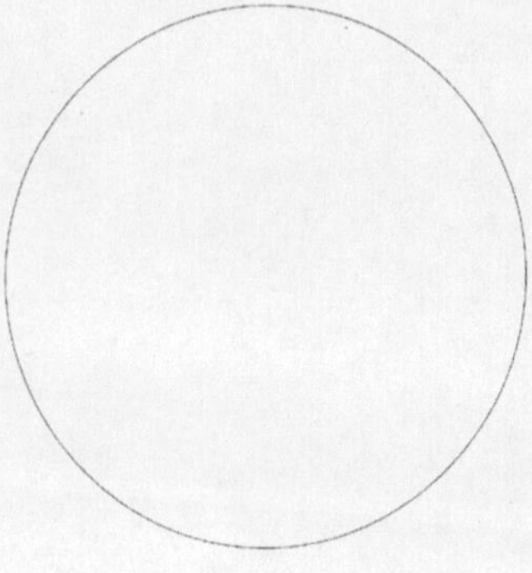

Montage by architect Ciro Miguel from the 2017 series *The Sunken Planets* showing Oscar Niemeyer's Ibirapuera Park exhibition building (1951-1954) augmented by a figure-ground silhouette of the Earth – a play on the Platonic idealism of architecture as well as on the notion of the planet itself as the exemplary model of spatial enclosure.[45] (courtesy Ciro Miguel, Department of Architecture, ETH Zurich)

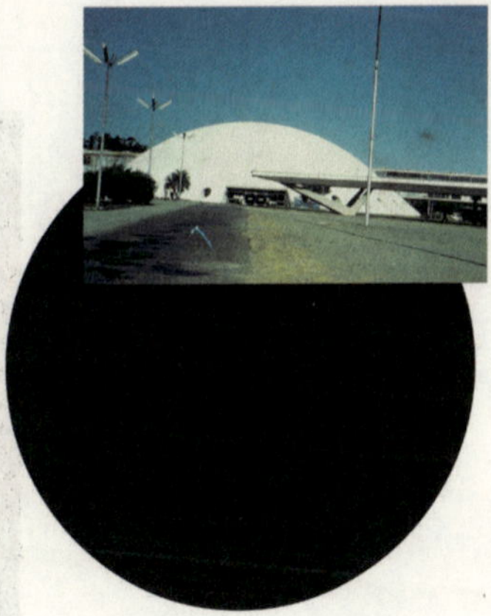

Animated impact model by David N. Bresch and Tarun Chadha from ETH Zurich showing the sum of tropical cyclones in the Asia-Pacific region from 1987 to 2016 as they develop and hit exposed land masses, the majority of which are highly urbanized; the model demonstrates a drastic increase in the frequency of such storms as well as their growing impact on coastal zones, 2016. (courtesy David N. Bresch and Tarun Chadha, Institute for Environmental Decisions, Department of Environmental Systems Science, ETH Zurich and MeteoSwiss)

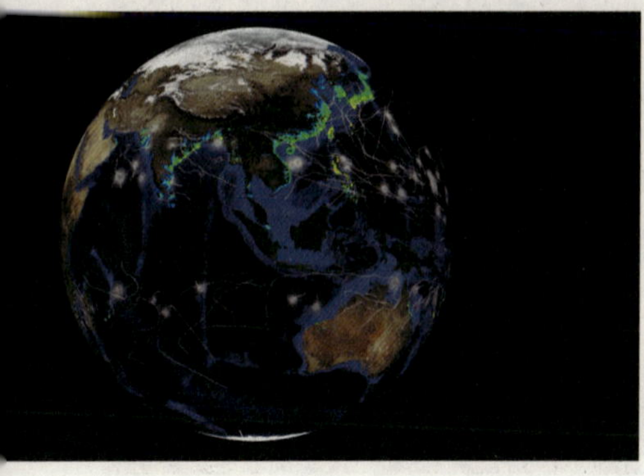

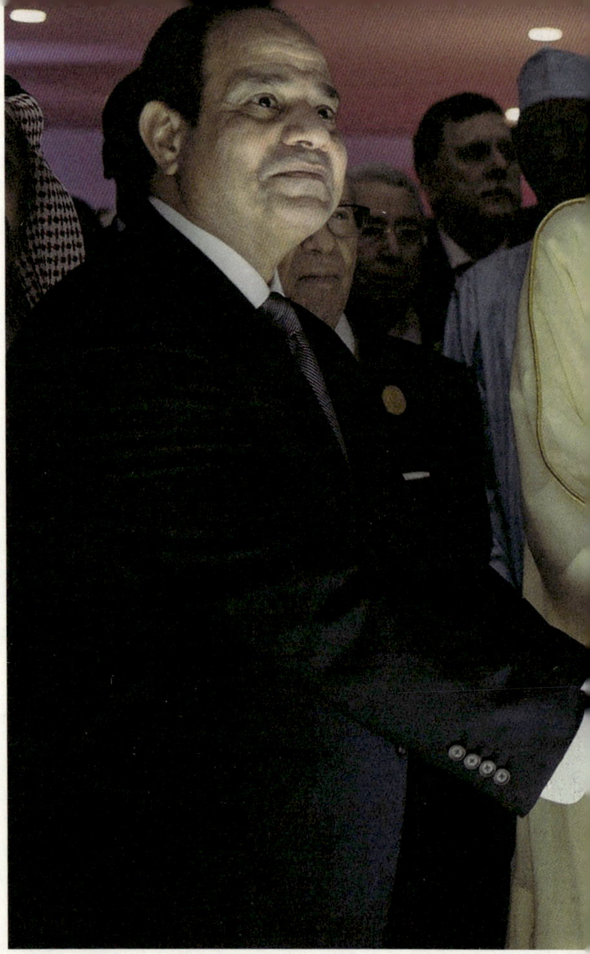

President Abdel Fattah el-Sisi of Egypt, King Salman of Saudi Arabia, and United States President Donald Trump seen with their hands on a glowing globe at the inauguration of the new Global Center for Combating Extremist Ideology based in Riyadh, Saudi Arabia, May 21, 2017. (courtesy Saudi Press Agency)

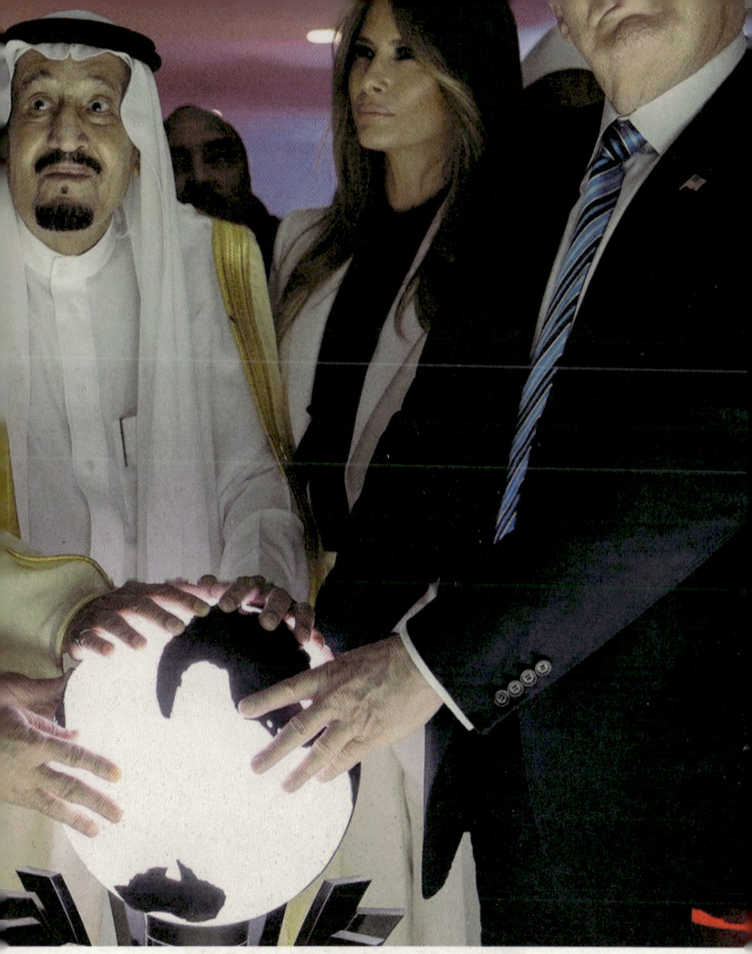

Unauthorized Greenpeace message reading "Planet Earth First" as projected on the dome of St. Peter's Basilica on the eve of Donald Trump's visit to the Vatican to meet Pope Francis in May 2017; just after the visit Trump announced at the G7 Summit in Sicily that the United States would consider withdrawing from the Paris Agreement on climate change. (courtesy Greenpeace and Getty Images)

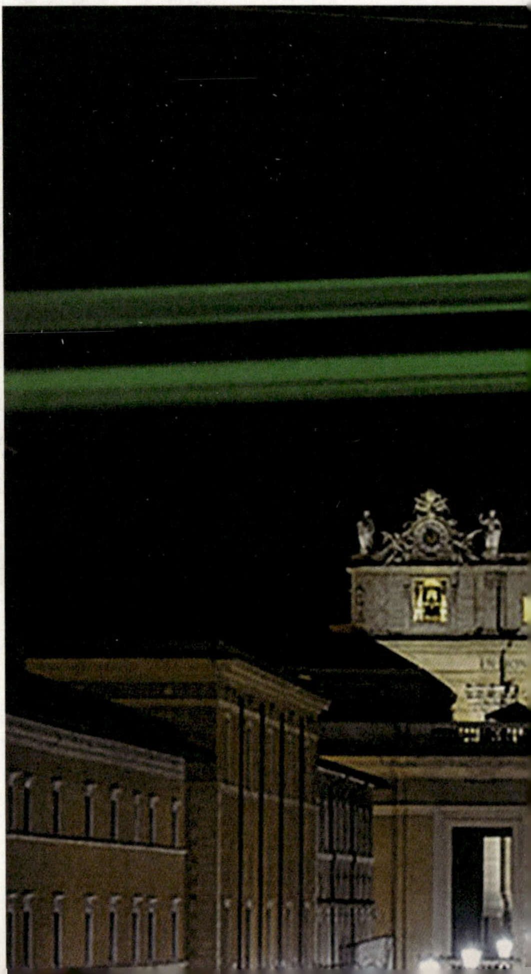

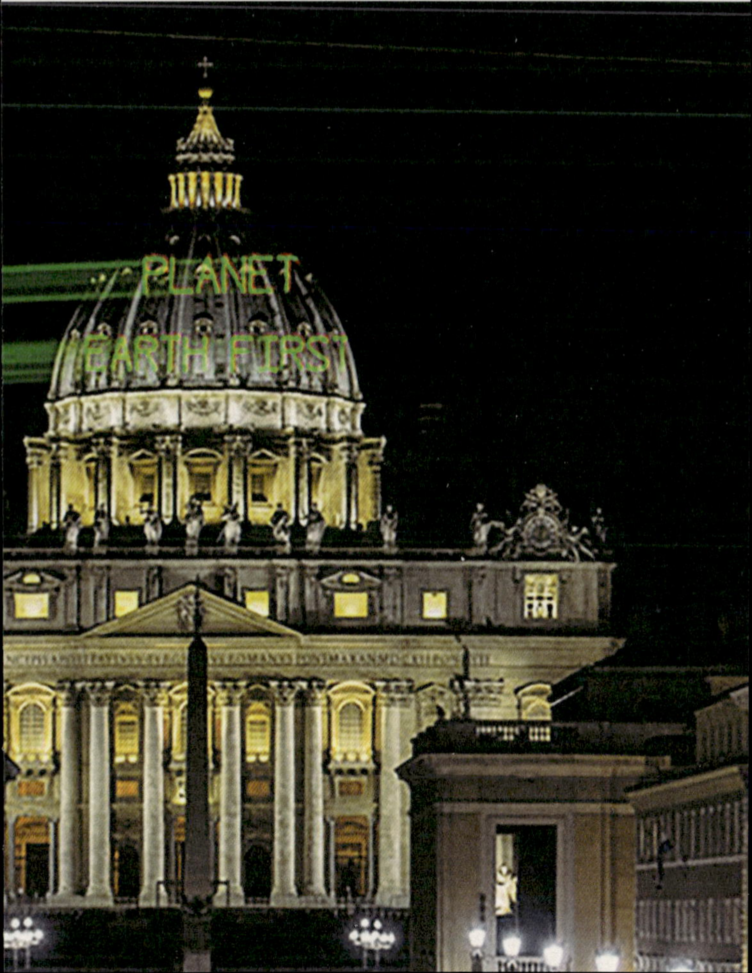
PLANET
EARTH FIRST

Cover of *The Economist* depicting Donald Trump as a "demolition man" gleefully astride the wrecking ball of the globe – a not so subtle commentary on his reckless foreign policy initiatives worldwide, June 9-15, 2018. (drawing by Ben Kirchner, courtesy The Economist)

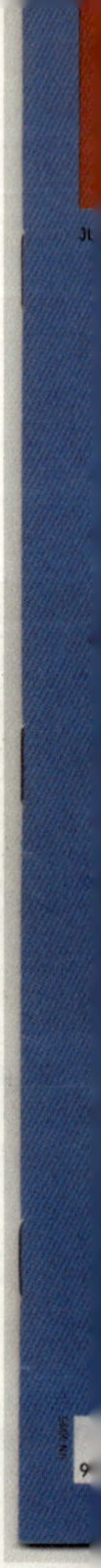

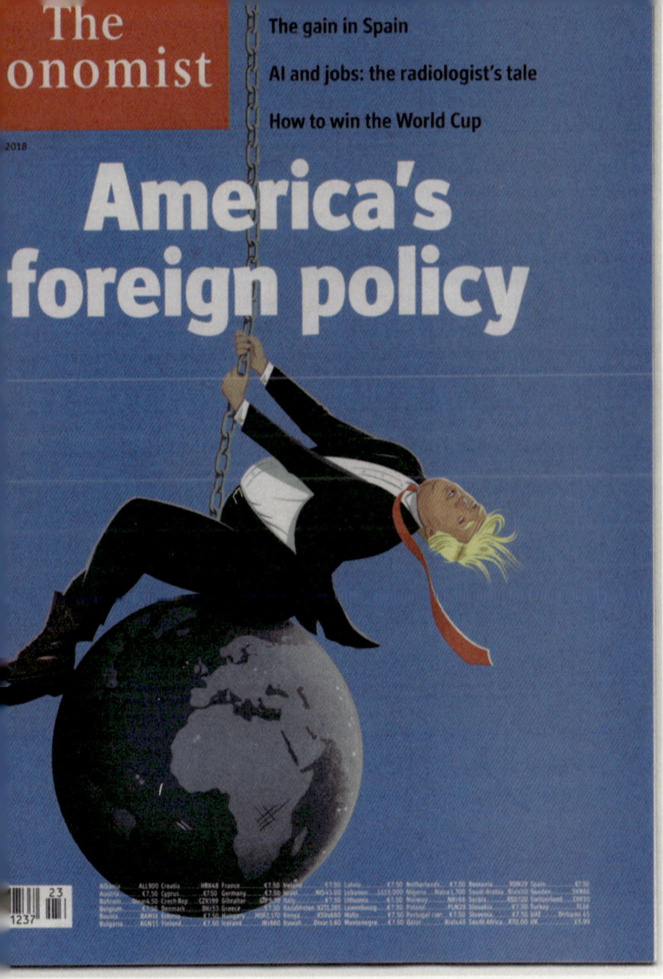
The
onomist
2018
The gain in Spain
AI and jobs: the radiologist's tale
How to win the World Cup
America's foreign policy

Still from a digital animation of seismic waves generated by the 2011 Tōhoku-Oki earthquake off the coast of Japan, triggering a major tsunami that in turn caused extensive damage, most notably at the Fukushima Nuclear Power Plant; the research required a numerical simulation with almost 8 billion degrees of freedom, which was carried out on the Swiss Piz Daint supercomputer and ran in near realtime on 1920 cores; the visualization was produced at the Institute of Geophysics, Department of Earth Sciences, ETH Zurich, supported by the Swiss National Supercomputing Centre (CSCS) and Mondaic AG, 2018-19. (courtesy Domenico Giardini, Lion Krischer, Michael Afanasiev, Christian Boehm, Martin van Driel, Andreas Fichtner, Martin van Driel)

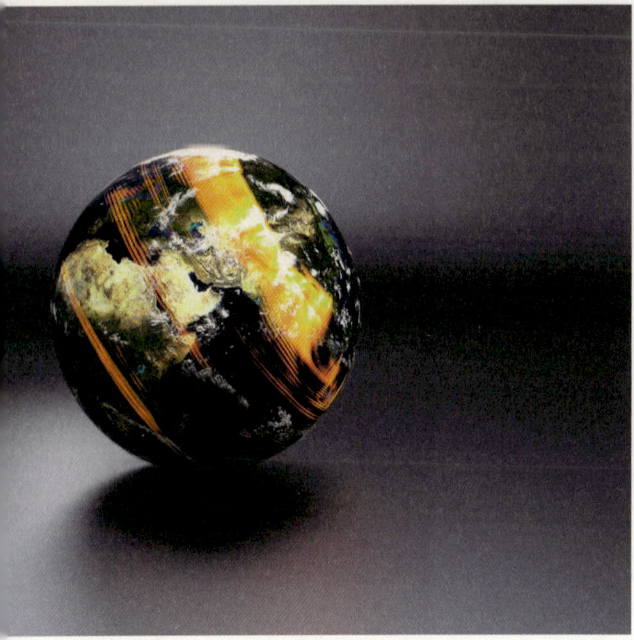

Diagrams showing investments from 2006-2015 by the top 10 Multilateral Development Banks (MDBs), such as the World Bank or European Investment Bank, in financing projects for electricity generation primarily in developing countries, whether fossil fuel-based or via hydropower and new renewables; this 'greening' of the technology portfolio of MDBs is allegedly aimed at reducing the climate-related risk exposure of the banks; Energy Politics Group, Department of Humanities, Social and Political Sciences, ETH Zurich, 2018. (figures Bjarne Steffen and Tobias Schmidt)

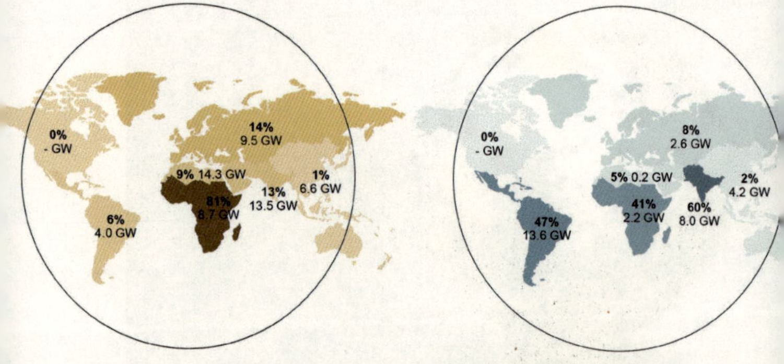

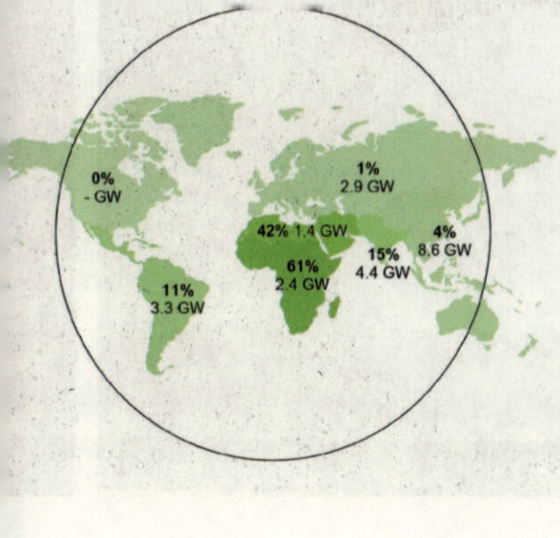
0%
- GW
1%
2.9 GW
42% 1.4 GW
4%
8.6 GW
15%
4.4 GW
61%
2.4 GW
11%
3.3 GW

Video stills of an atmospheric event known as Sudden Stratospheric Warming (SSW) that occurred in mid-February 2018, whereby the polar vortex split into two separate vortices, in turn affecting weather on the Earth's surface; in particular, such events can lead to persistent cold spells over specific world regions; a similar event occurred in early 2019; Institute of Atmospheric and Climate Science, Department of Environmental Sciences, ETH Zurich and Hochschule Düsseldorf, 2018.[46] (courtesy Daniela Domeisen and Alexander Wollert)

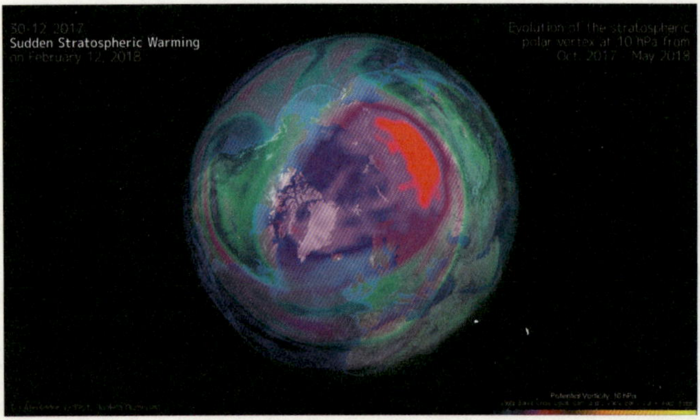

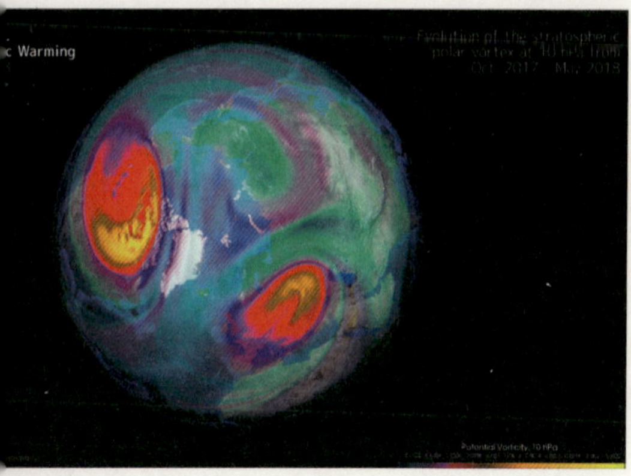
c Warming
Evolution of the stratospheric
polar vortex at 10 hPa from
Oct. 2017 - May 2018
Potential Vorticity, 10 hPa

Interactive virtual model produced through photogrammetric processing of multiple images of the original and replica of the *St. Galler Globus*, located in the Landesmuseum in Zurich and Stiftsbibliothek St. Gallen, respectively, 2018-19.[47] (Konrad Schindler et al., Photogrammetry and Remote Sensing, Institute of Cartography and Geoinformation, Department of Civil, Environmental and Geomatics Engineering, ETH Zurich and Landesmuseum Zürich)

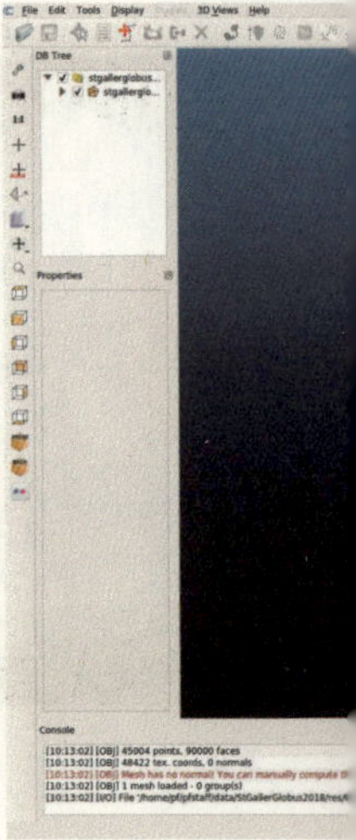

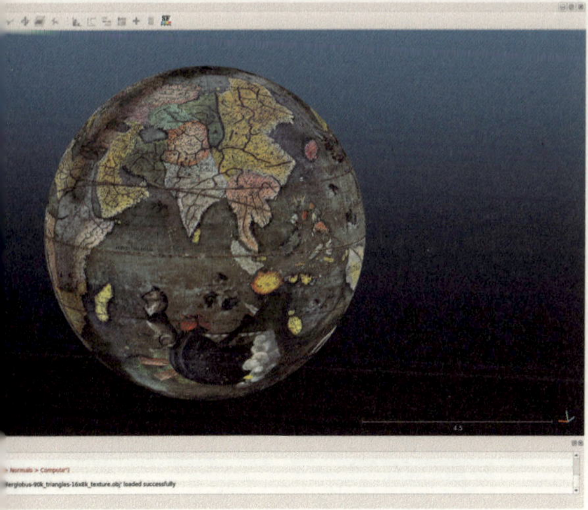
> Normals > Compute']
erglobus-90k_triangles-16x8k_texture.obj' loaded successfully

Composite territories of Planetary Urbanization in the second decade of the 21st century by Nikos Katsikis, based on global geospatial datasets that show agglomeration zones (blue gradient) and operational landscapes of production and circulation (red gradient). (cartography and modeling by Nikos Katsikis, 2018)

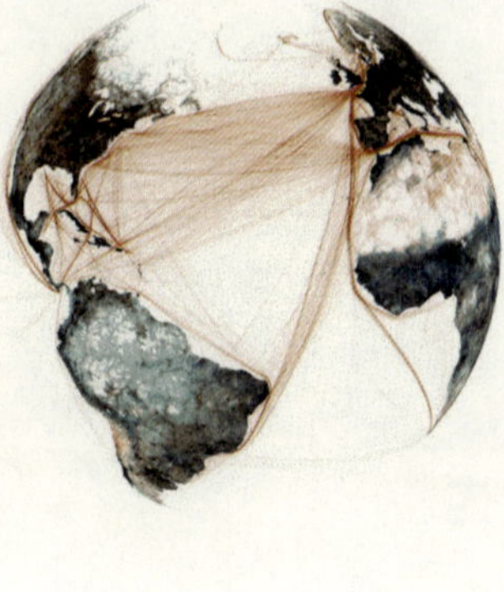

CNC model of the equipotential surface of the Earth's gravity field forming what is called a "geoid," 2019.[48] (courtesy Alain Geiger, Institute of Cartography and Geoinformation, Department of Civil, Environmental and Geomatics Engineering, ETH Zurich)

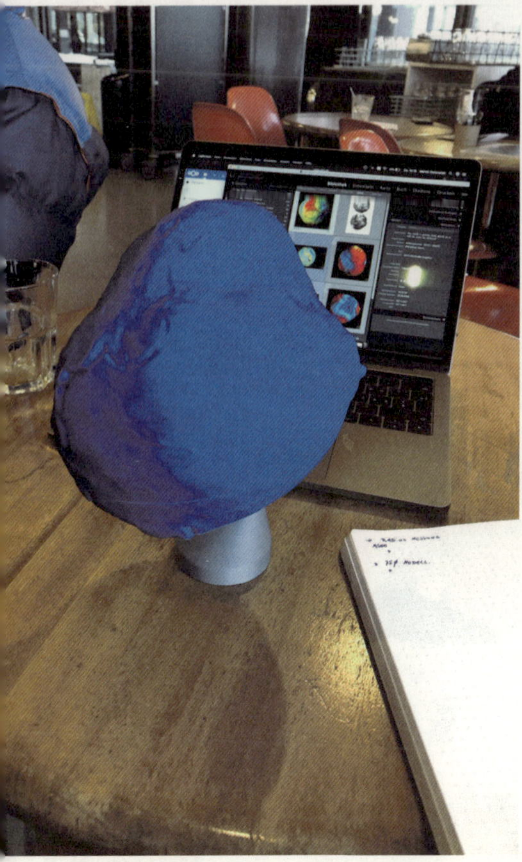

Image entitled "The Changing Ocean," which shows the evolution of observed CO_2 concentrations on the surface of the world's oceans from 1982 to 2017 that has resulted in a drastic rise in water temperature; data from Landschützer et al., 2016; figure prepared by Nicolas Gruber, 2019.[49] (courtesy Nicolas Gruber, Environmental Physics Group, Institute of Biogeochemistry and Pollutant Dynamics, ETH Zurich)

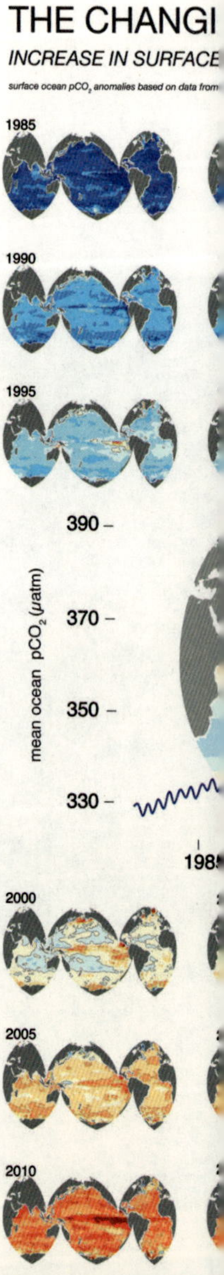

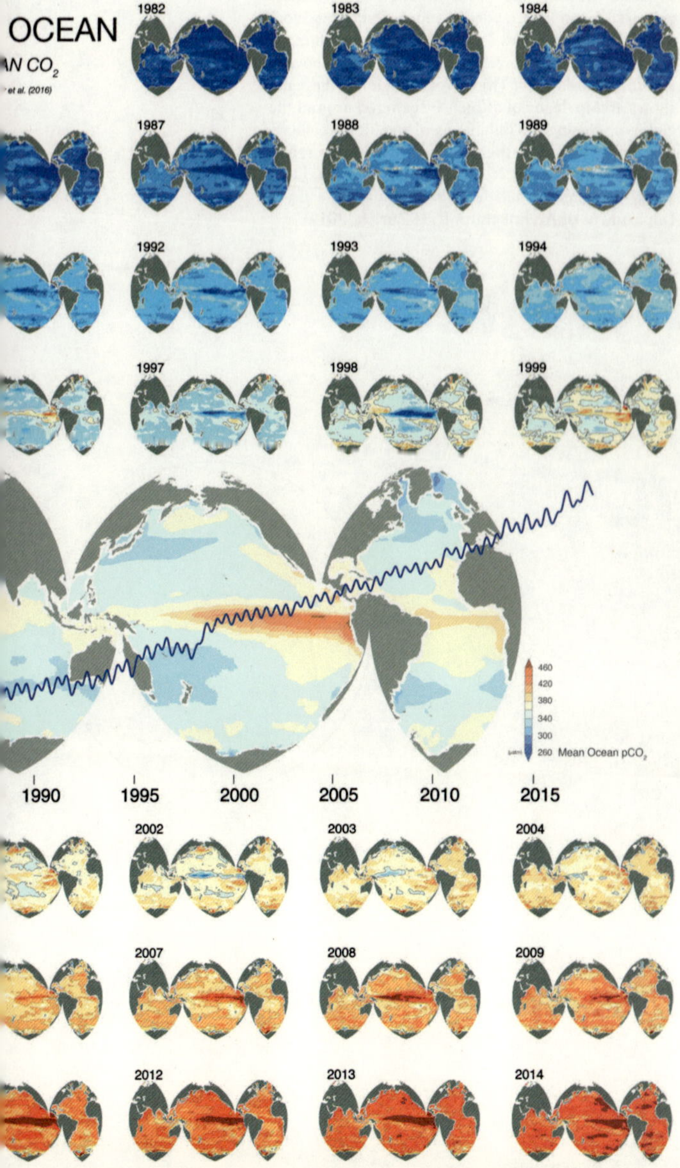
OCEAN
AN CO2
et al. (2016)
1982
1983
1984
1987
1988
1989
1992
1993
1994
1997
1998
1999
460
420
380
340
300
260 Mean Ocean pCO2
1990
1995
2000
2005
2010
2015
2002
2003
2004
2007
2008
2009
2012
2013
2014

Diagrams of uneven world development; the first is the so-called "Brandt Line" from the 1980 *Brandt Report* prepared by an independent commission of the World Bank, proposing a new "poverty line" demarcating what was then called the developed "First World" above and the underdeveloped "Third World" below;" the second shows archipelagos of affluence scattered around the globe, disclosing the imbalance of wealth accumulation worldwide, based on the Brookings Institution report *Global MetroMonitor 2014*.[50] (Leonard Streich, Elena Schütz, Julian Schubert from Something Fantastic, Department of Architecture, ETH Zurich, 2019)

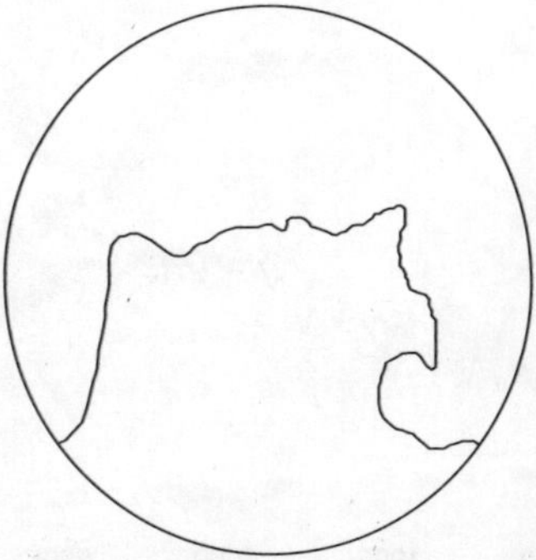

[illegible]

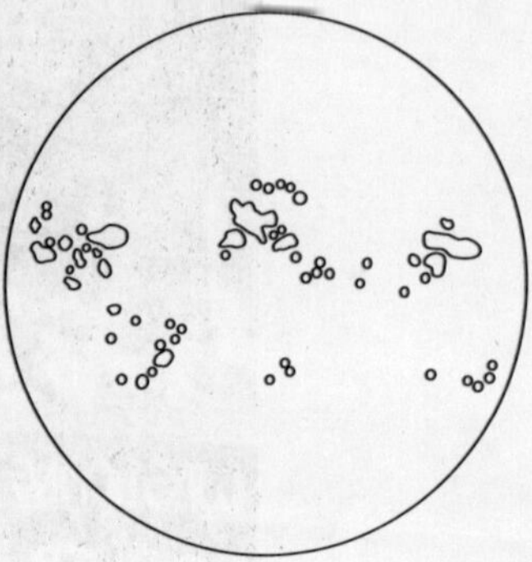

One of the many "Fridays for Future" protests organized by youth around the globe to prompt responsive action on climate change, the mock globe being volleyed above their heads perhaps meant to signal that the state of the world has gotten out of hand; pupils and students are calling on leaders to back up conciliatory rhetoric with accountable deeds, making the original definition of sustainable development from the 1987 *Brundtland Report* all the more timely as that which "meets the needs of the present without compromising the ability of future generations to meet their own needs," Sydney, March 14, 2019. (courtesy The New York Times and Reuters)

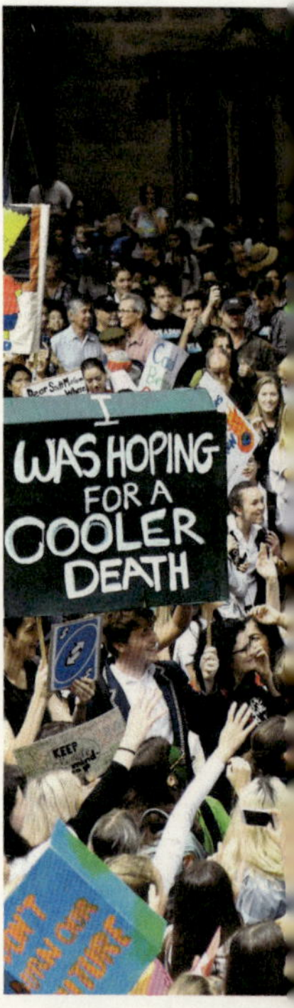

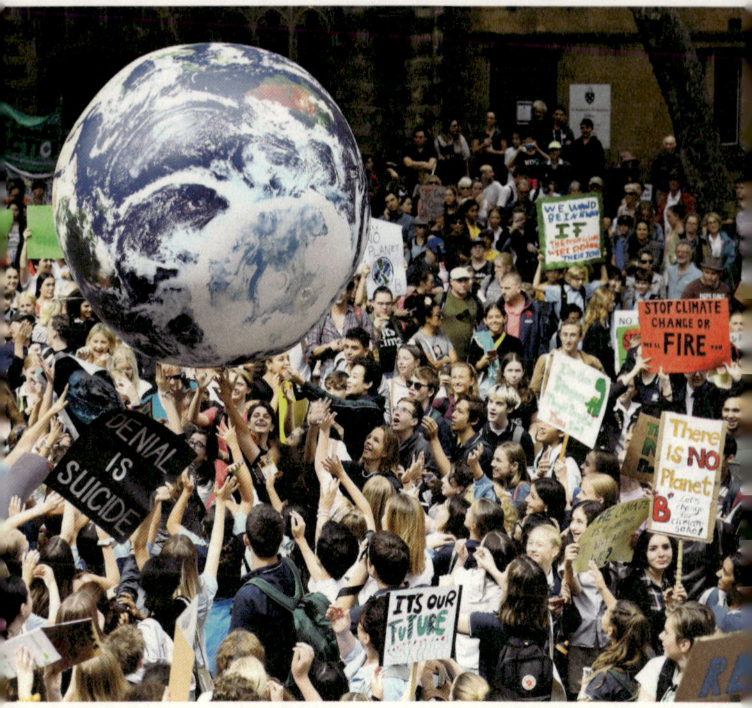
DENIAL IS SUICIDE
STOP CLIMATE CHANGE OR WE'LL FIRE ...
There is NO Planet B
ITS OUR FUTURE

1 Gilles Deleuze and Félix Guattari, *A Thousand Plateaus: Capitalism and Schizophrenia*, trans. Brian Massumi (Minneapolis: University of Minnesota Press, 1987), 39; originally published as *Mille Plateaux* (Paris: Les Éditions de Minuit), 1980.

2 John Lowden, *The Making of the Bibles Moralisées—Volume I: The Manuscripts* (University Park, PA: The Pennsylvania State University Press, 2000), 8-94.

3 Incidentally, the image by Andy Sacks was also used for the cover of a book on the effects of globalization on labor, see Kimberly Ann Elliott and Richard B. Freeman, *Can Labor Standards Improve Under Globalization?* (Washington, DC: Institute for International Economics, 2003).

4 See Michael Marshall, "The Universe Is an Egg and the Moon Isn't Real: Notes from a Flat Earth Conference," *The Guardian*, May 2, 2018.

5 Joseph Kahn, "Storm Over 1418 Map: History or Scam?", *The New York Times*, January 16, 2006.

6 See Buckminster Fuller and John McHale, "Phase I—Document 1: Inventory of World Resources, Human Trends, and Needs," report submitted to the International Union of Architects (IUA), Southern Illinois University, 1963 (www.bfi.org).

7 For the reference to Umberto Eco's concept of the "open work" and its significance for Karin Sander's installation, see Roman Kurzmeyer, "Karin Sander," *Atelier Amdem* (Zurich: Edition Voldemeer Zürich, 2015), 215-224.

8 Roman Kurzmeyer, "Zeit der Transformation," *Neue Zürcher Zeitung*, July 27, 2015.

9 For the terms "geostory" and "common geostory," see Bruno Latour, "Facing Gaia: Six Lectures on the Political Theology of Nature," unpublished manuscript for Latour's Gifford Lectures on Natural Religion, Edinburgh, February 18-28, 2013. See also Bruno Latour, "Agency at the Time of the Anthropocene," *New Literary History* 45 (2014): 1-18; lecture transcript prepared for the Holberg Prize Symposium 2013: "From Economics to Ecology," Bergen, June 4, 2013.

10 For the term "geostories" see the publication by Design Earth: Rania Ghosn and El Hadi Jazairy, *Geostories: Another Architecture for the Environment* (Barcelona: Actar, 2018).

11 Ghosn and Jazairy, *Geostories*, 12.

12 Martin Heidegger, "The Age of the World Picture," *Off the Beaten Track,* Julian Young and Kenneth Haynes, eds. and trans. (Cambridge: Cambridge University Press, 2002), 57-85; originally published as "Die Zeit des Weltbildes" (1938), *Holzwege* (1935–46), *Gesamtausgabe,* Band 5 (Frankfurt am Main: Vittorio Klostermann, 1977), 87-94.

13 Thomas L. Friedman, *The World is Flat: A Brief History of the Twenty-First Century* (New York: Farrar, Strauss and Giroux, 2005).

14 Peter Sloterdijk, *Globes—Spheres Volume II: Macrospherology*, trans. Wieland Hoban (New York: Semiotext(e), 2014), Chapter 8, "The Last Orb: On a Philosophical History of Terrestrial Globalization," 765-967; originally published as *Sphären II—Globen: Makrosphärologie* (Frankfurt am Main: Suhrkamp Verlag).

15 Sloterdijk, *Globes*, 25; see also Peter Sloterdijk, "The Time of the Crime of the Monstrous: On the Philosophical Justification of the Artificial," *Sloterdijk Now*, ed. Stuart Elden (Cambridge: Polity Press, 2012), 169.

16 Sloterdijk, "The Time of the Crime of the Monstrous," 168.

17 Peter Sloterdijk, *In the World Interior of Capital*, trans. Wieland Hoban (Cambridge: Polity Press, 2013); originally published as *Im Weltinnenraum des Kapitals: Für eine philosophische Theorie der Globalisierung* (Frankfurt am Main: Suhrkamp Verlag, 2005); see also Chapter 8 of *Globes – Spheres Volume II*.

18 Sloterdijk, *In the World Interior of Capital*, 171.

19 Bruno Latour, *Inside*, performance lecture staged by Frédérique Ait-Touati, with cartographic diagrams by Alexandra Arènes, Sonial Lévy, and Axelle Grégoire, presented at the HAU Theater in Berlin, September 20, 2017; see www-bruno-latour.fr.

20 See Bruno Latour, *Down to Earth: Politics in the New Climatic Regime*, trans. Catherine Porter (Cambridge: Polity Press, 2018); originally published as *Où atterir? Comment s'orienter en politique* (Paris: Éditions la Découverte, 2017).

21 Though originally coined by biologist Eugene Stoermer in the early 1980s, the concept of the Anthropocene gained notoriety when introduced by atmospheric chemist Paul Crutzen in 2000 to denote the succeeding geological era after the Holocene, see Paul J. Crutzen and Eugene F. Stoermer, "The 'Anthropocene'," *IGBP Newsletter* 41 (May 2000): 17–18.

22 Will Steffen, Paul J. Crutzen, and John R. McNeill, "The Anthropocene: Are Humans Now Overwhelming the Great Forces of Nature?," *Ambio* 36, no. 8 (December 2007): 614–621.

23 Erik Swyngedouw, "More-Than-Human Constellations as Immuno-biopolitical Fantasy in the Urbicene," *New Geographies* 9 (January 2018): 19–23.

24 Neil Brenner, "Introduction: Urban Theory without an Outside," *Implosions / Explosions*, ed. Neil Brenner (Berlin: Jovis Verlag, 2014), 18.

25 Jason W. Moore, *Capitalism in the Web of Life: Ecology and the Accumulation of Capital* (London: Verso, 2015), see chapter entitled "Anthropocene or Capitalocene?," 169-192.

26 Moore, *Capitalism in the Web of Life*, 174.

27 See Raj Patel and Jason W. Moore, *A History of the World in Seven Cheap Things: A Guide to Capitalism, Nature, and the Future of the Planet* (Oakland: University of California Press, 2017).

28 Neil Smith, *Uneven Development: Nature, Capital, and the Production of Space* (Athens, GA: University of Georgia Press, 1984); see chapter entitled "The Production of Nature" from the 2008 edition, 41-91.

1 Diagrammatic sections of the Pantheon: one showing the Roman temple as realized, the other the ideal geometry of a circle around which the interior volume was conceived.

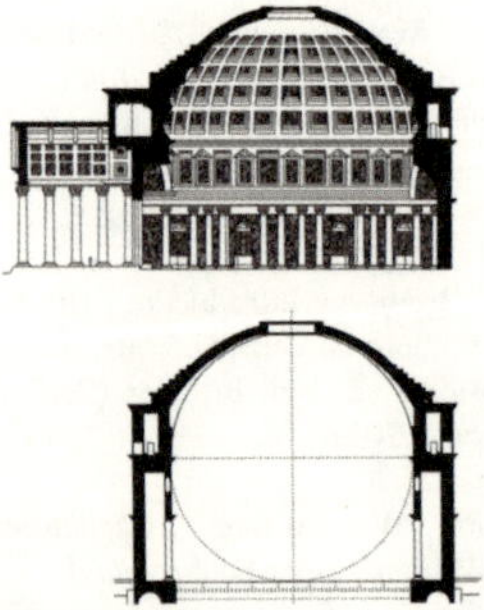

2 Facsimile of the Behaim-Globe as an unfolded surface produced by Ernest George Ravenstein (1908) and Armin Grün's digital visualization of the "Erdapfel," ETH Zurich (2011-12).

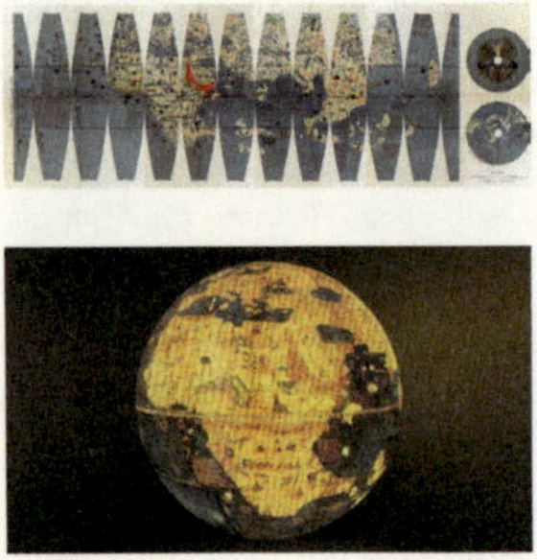

3 Interior panels of Hieronymus Bosch's triptych *The Garden of Earthly Delights* thematizing terrestrial life in an amalgam of imaginary scenes of vices and virtues that transcend reality, ca. 1505 (courtesy Museo del Prado and © akg images).

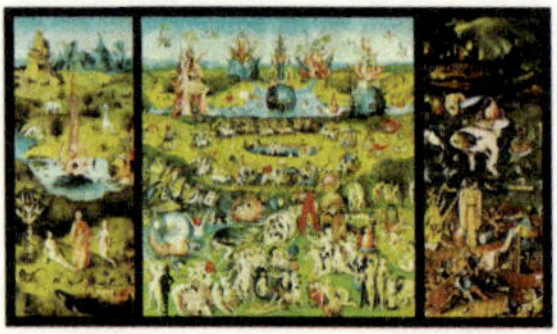

4 Parchment by Peter Boeckel from 1595 depicting the St. Galler Erd- und Himmelsglobus discovered in the Brockenhaus Olten in 2016 (Zentralbibliothek Zürich).

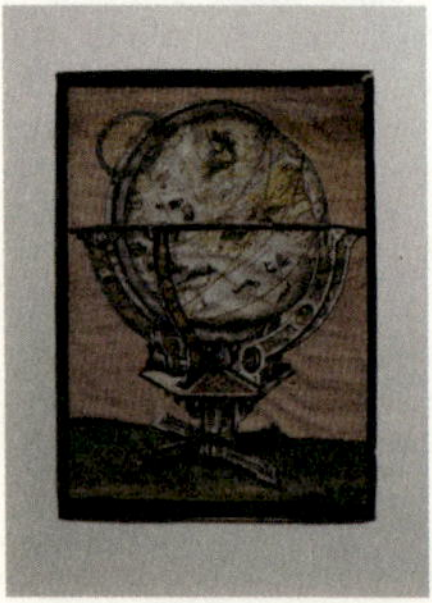

5 The Woburn Abbey version of the *Armada Portrait* of Queen Elizabeth I showing her sitting in front of two views commemorating the defeat of the Spanish fleet, attributed to George Gower, 1588 (courtesy Woburn Abbey and © akg images).

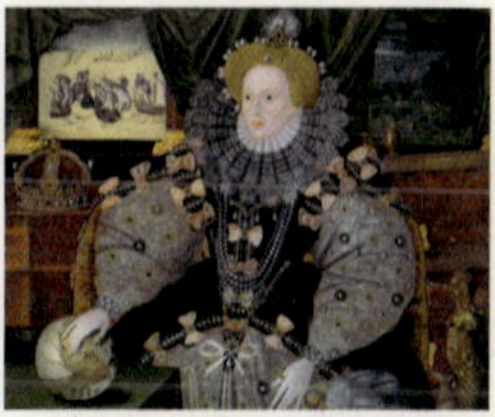

6 Detail of fresco by Giulio Parigi in the hall dedicated to the science of mathematics in the Uffizi Gallery in Florence from 1599-1600 and engraving from the cover of *Mechanics Magazine* illustrating the lever principle of Archimedes, vol. 2, 1824.

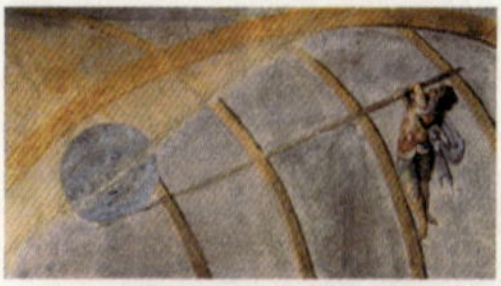

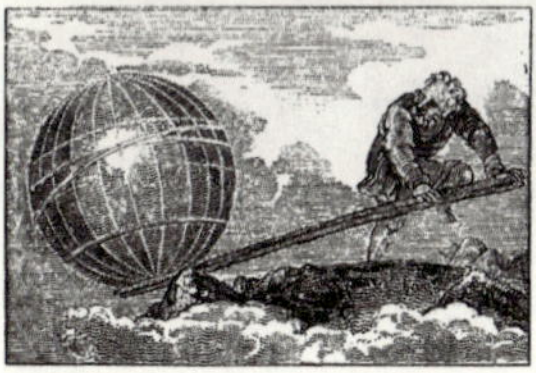

7 Study for a Goblin tapestry by Henri Testelin from 1667, with Louis XIV and Jean-Baptiste Colbert greeting members of the newly founded Royal Academy of Sciences (courtesy Musée national du Château de Versailles).

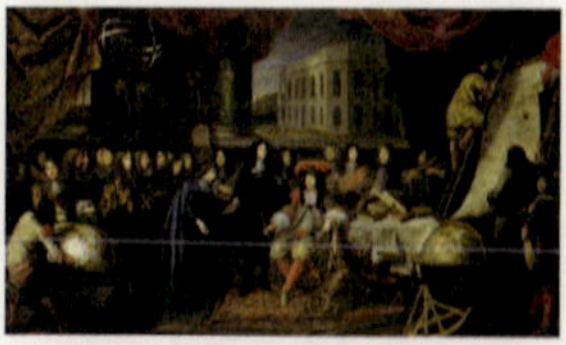

8 Drawing entitled *Élevation du cimetière de la ville de Chaux* showing a conceptual elevation view of the cemetery complex as a constellation of planetary bodies in space; from Claude-Nicolas Ledoux's 1804 treatise *L'architecture considérée sous le rapport de l'art, des moeurs et de la législation* (courtesy Bibliothèque Nationale de France).

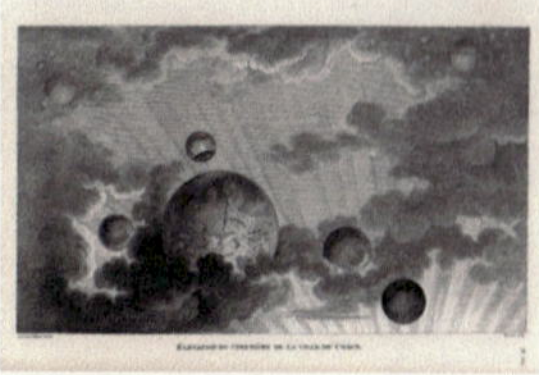

9 Drawing over a photograph showing Alberto Palacio's Chicago World Fair scheme for the *Monument in Memory of Christopher Columbus* resituated in Paris to make the point that it would be taller than the recently realized Eiffel Tower, ca. 1890.

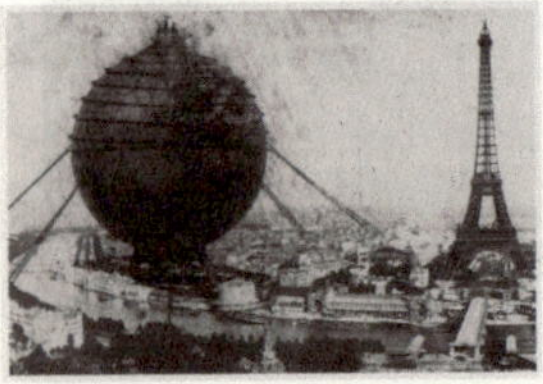

10 Relief of Switzerland by geographer Charles Eugène Perron made for Élisée Reclus's *Globe terrestre* for the Exposition universelle of 1900; the relief is curved as it had to be mounted on the inner sphere of Reclus's structure (Relief Collection of ETH Zurich, *focus*Terra).

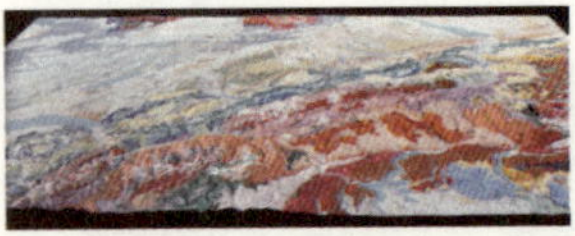

11 Albert Galeron's drawing for the *Cosmorama* at the Exposition universelle of 1900, a drawing made at the same time as geographer Élisée Reclus presented his *Globe terrestre* to the Sixth International Geographical Congress that convened in London in 1895 (courtesy Archives nationales, France).

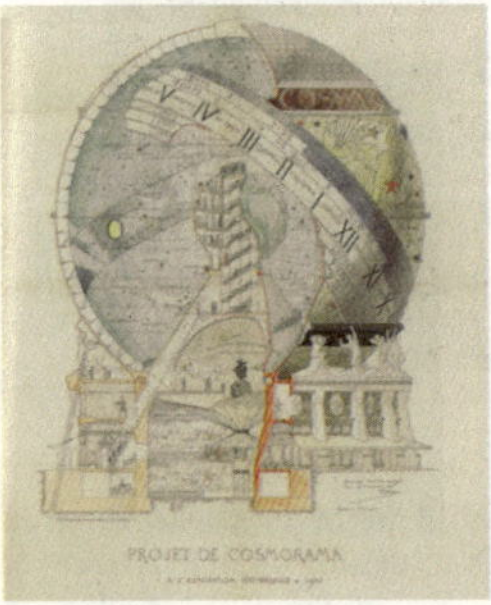

12 Photographic portrait of Theodore Roosevelt in 1909 near the end of his second term as president, again in the presence of the large globe that sat in the White House (photograph Rockwood Photo Co., courtesy Library of Congress)

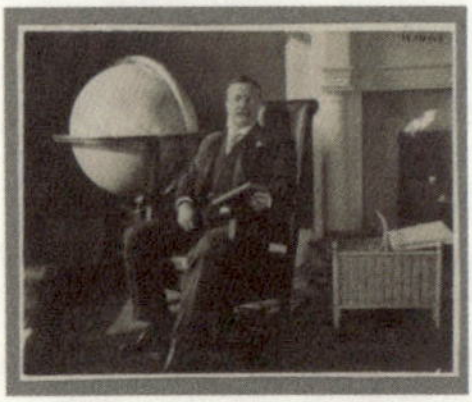

13 Model of thesis project by Ivan Illich Leonidov for the Lenin Institute for Librarianship, 1927 (courtesy The Museum of Modern Art, New York).

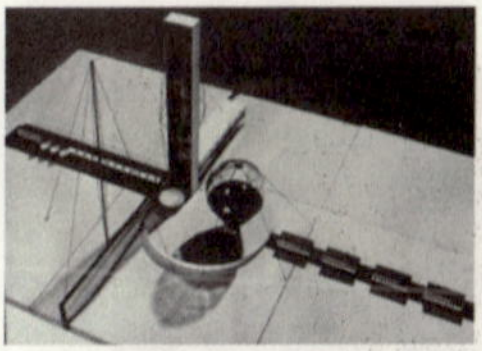

14 Sketch by Buckminster Fuller from his 1928 investigations on dwellings, in which he envisioned a world equipped with structures so light in weight that they could be transported by airships circling the globe (courtesy Buckminster Fuller Institute).

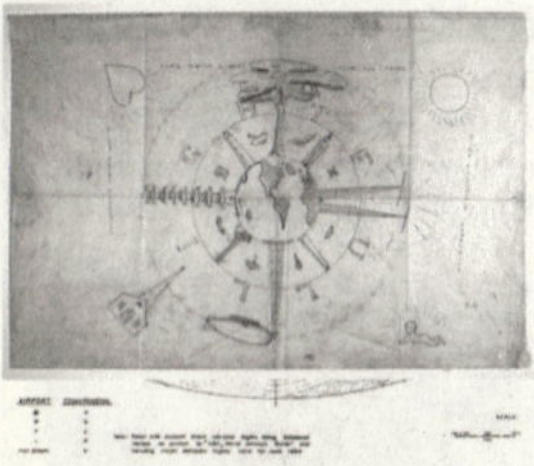

15 Photograph showing Buckminster Fuller with his Dymaxion World Map unfolded on the wall, *Life Magazine*, March 1943.

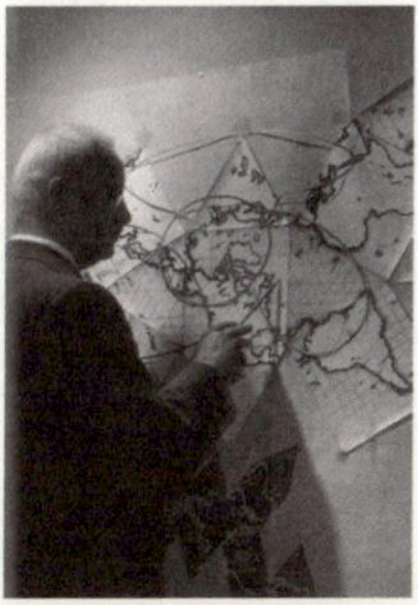

16 Poster by Joseph Binder and a limited-edition stamp from the United States Postal Service showing the iconic Perisphere and Trylon at the 1939 New York World's Fair as emblems of "The World of Tomorrow."

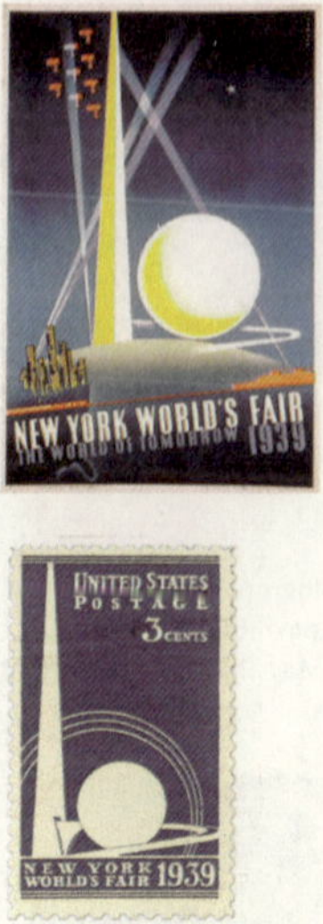

17 Drawings of the Pentadome showing the arrangement of the inflated semi-spherical missile hangers at Andrews Air Force Base, 1958-59 (courtesy United States Army).

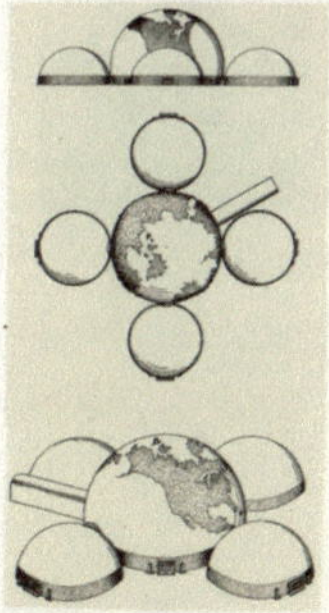

18 Part of Ecumenopolis (Constantinos Doxiadis's concept of the globally networked city) would include the infrastructural nodes servicing planetary air traffic as depicted in one of the diagrams in the journal *Ekistics* from July 1966 (courtesy Constantinos A. Doxiadis Archives / Constantinos and Emma Doxiadis Foundation).

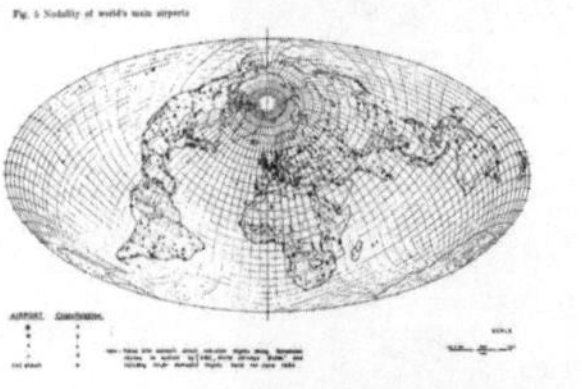

19 Aerial photograph of the 1967 Montreal Expo pavilion by Buckminster Fuller on fire, May 20, 1976 (courtesy Associated Press)

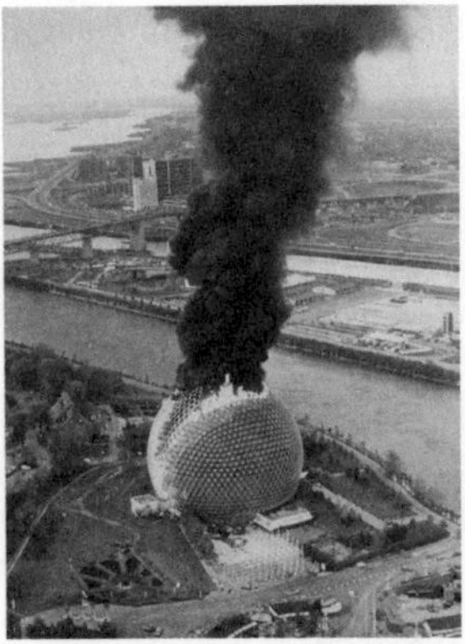

20 Poster announcing the documentary film *Earthrise* by Emmanuel Vaughan-Lee, released in April 2018 (courtesy Adam Loften and Emmanuel Vaughan-Lee).

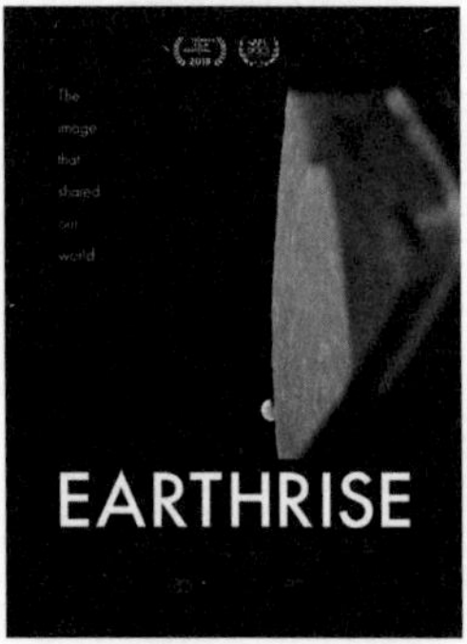

21 Buckminster Fuller conducting the first *World Game* seminar at the New York Studio School for Drawing, Painting, and Sculpture, 1969 (courtesy Buckminster Fuller Institute).

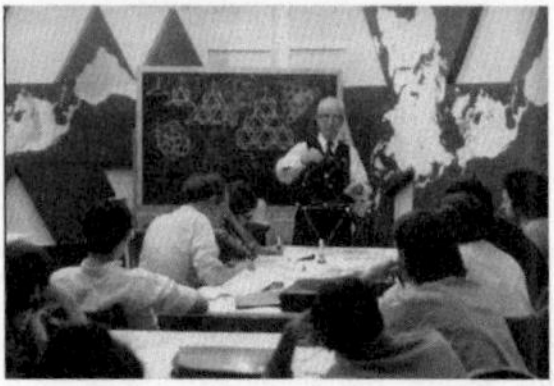

22 Photograph of activist Peter Hallerman with a gas mask sniffing magnolia blossoms during the first Earth Day demonstration in New York City on April 22, 1970, a press image that struck a nerve around the country at the time when environmental concerns were gaining ground (courtesy Associated Press).

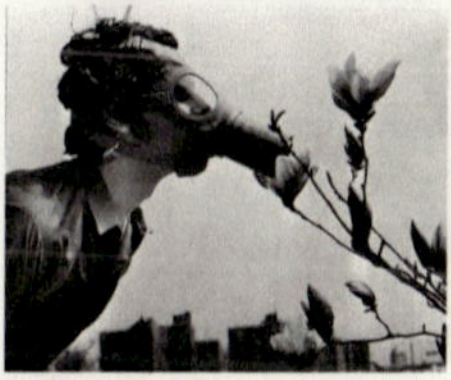

23 Back cover of the foreign edition of *The Last Whole Earth Catalog* from 1971, showing the "Earthrise" image taken by the crew of Apollo 8 in 1968 (submitted by Stephen Cairns, Future Cities Laboratory Singapore, Department of Architecture, ETH Zurich), and the 1967 ATS-3 satellite composite of planet Earth (courtesy National Aeronautics and Space Administration).

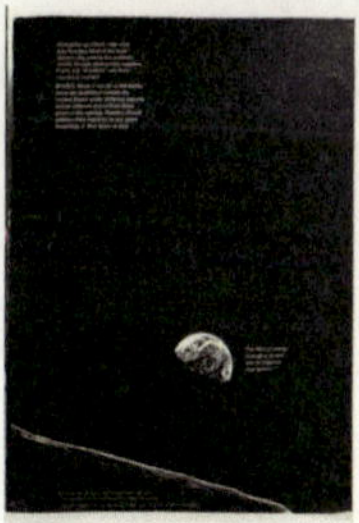

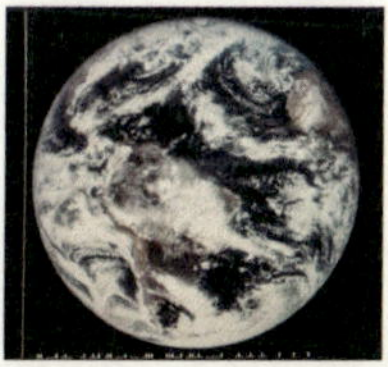

24 *Whole Earth Field Guide,* ed. Caroline Maniaque-Benton with Meredith Gaglio (Cambridge, Mass.: MIT Press, 2016).

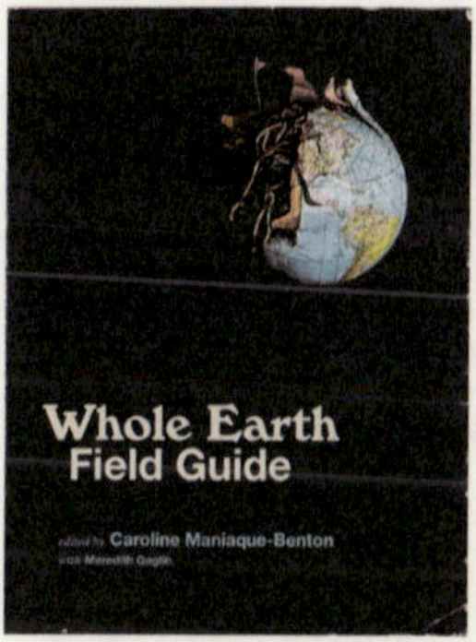

25 In later editions of *The Limits to Growth*, the 'green' Earth icon used on the original cover was replaced by the image of Earth in shackles and, in the German edition, Earth being hatched from an egg.

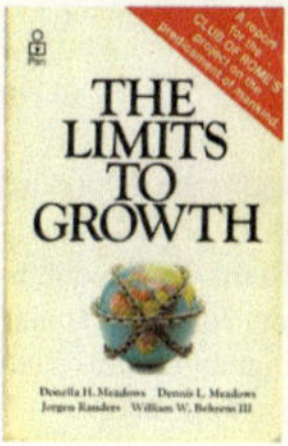

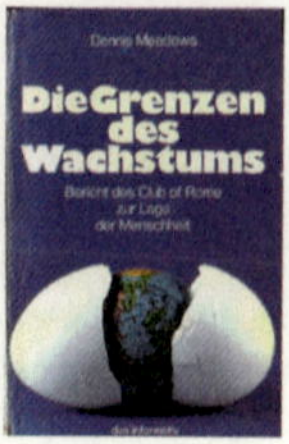

26 Still from Petra Noordkamp's film *La madre, il figlio e l'architetto*, 2012.

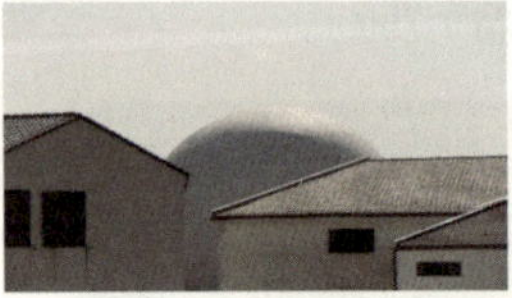

27 Artwork by Javacheff Christo entitled *Wrapped Globe*; project for the Institute of Contemporary Art, University of Pennsylvania, Philadephia, 1968.

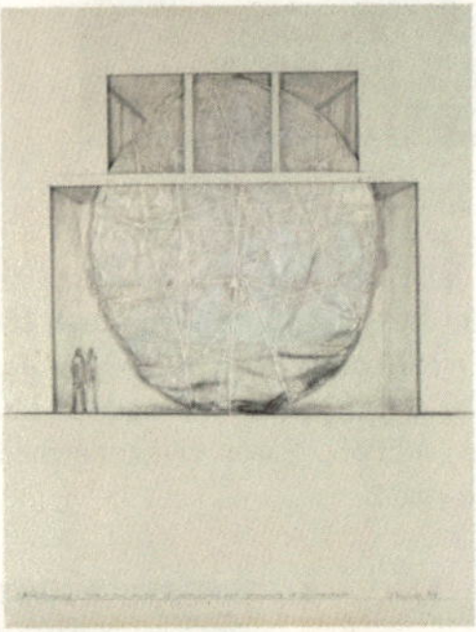

28 Cover of a book on the effects of globalization on labor, see Kimberly Ann Elliott and Richard B. Freeman, *Can Labor Standards Improve Under Globalization?* (Washington, DC: Institute for International Economics, 2003).

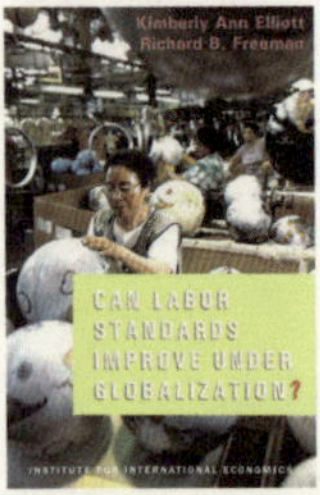

29 A forerunner to Google Earth was NASA's *Worldwind* virtual globe software, an alternative geospatial application released in 2003 for use on personal computers and as an open source platform in 2004 (courtesy National Aeronautics and Space Administration).

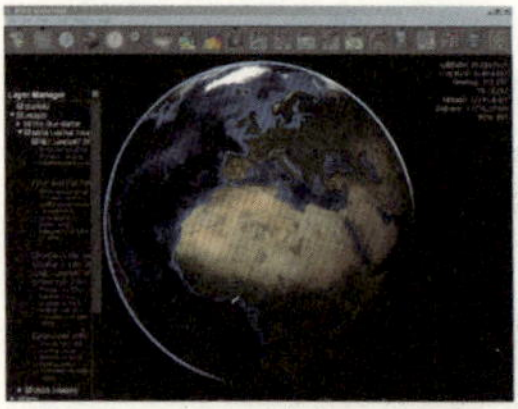

30 Sketch by Philippe Rekacewicz entitled *La dernière carte Nord-Sud: Esquisse finalisée* showing global flows of capital among world regions, 2002–3 (courtesy Philippe Rekacewicz, visionscarto.net)

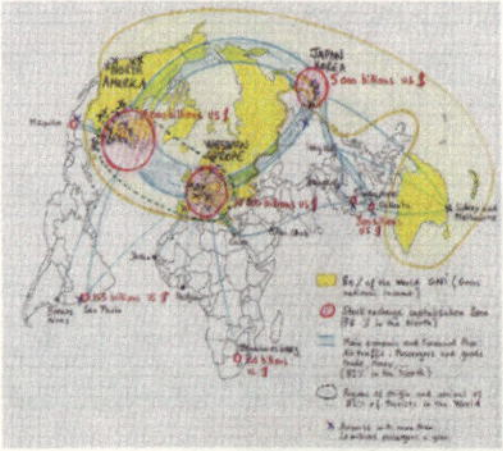

31 *Endlessly Open Cities* by Alexander Lehnerer, ETH Zurich, in collaboration with the University of Illinois at Chicago, 2009-10: *City of the Adjacent Cul-de-sac*, Matthew Vander Ploeg; *City of the Invasive Flora*, Liliana Aguirre.

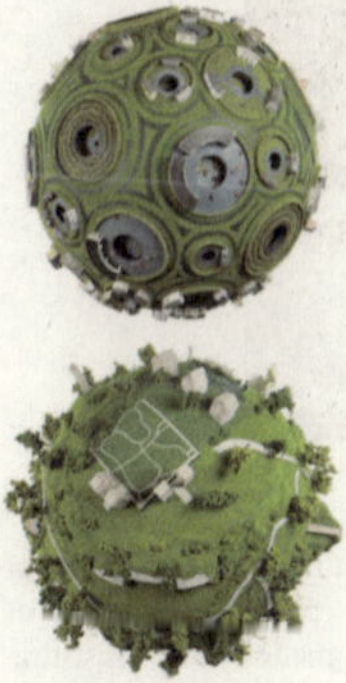

32 Model of *The World* displayed in the showroom of Nakheel Properties, Dubai, 2010 (photograph Marc Angélil)

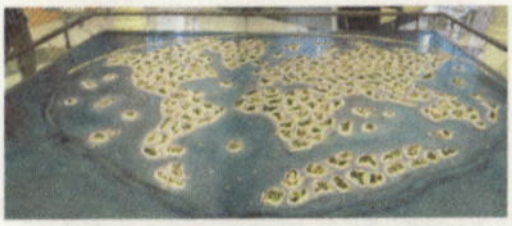

33 Cover of *The Energy Report: 100% Renewable Energy by 2050* from 2011, compiled by WWF International in collaboration with Ecofys and OMA/AMO, the latter under the supervision of Reinier de Graaf and Rem Koolhaas.

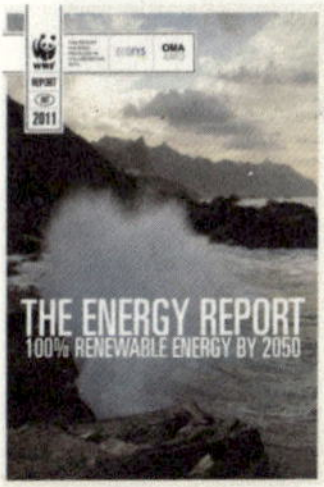

34 Cover of *The Economist* entitled "Welcome to the Anthropocene: Geology's New Age," May 28 - June 3, 2011; illustration by Jon Berkeley.

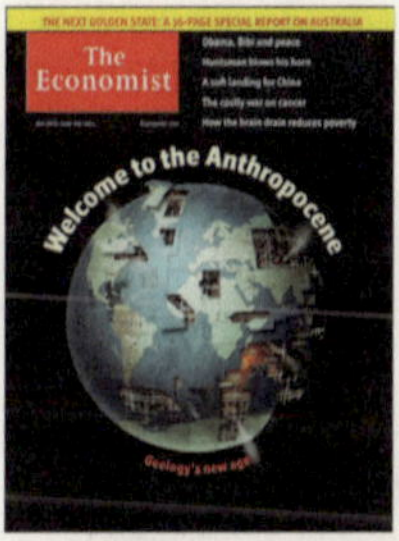

35 Bachelor thesis projects by Christian Lorenz (2011) and Christian Peier (2012) exploring the technical potential of using Tactile Hyperglobes and OmniGlobes for the visualization of global processes on spherical displays; research conducted at the Institute of Cartography and Geoinformation under the supervision of Lorenz Hurni, ETH Zurich.

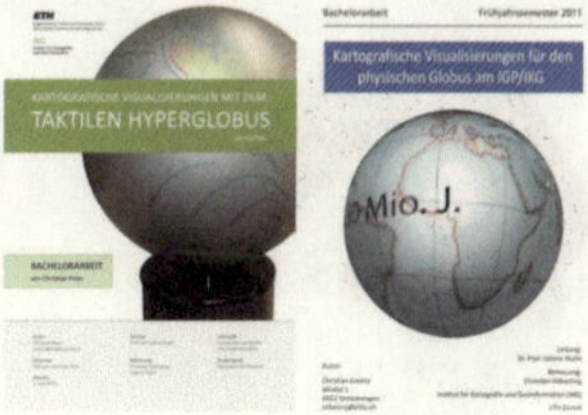

36 Rendering by Olivier Mestre for the 2014 version of *luna*, an inflatable structure by the Atelier de la conception de l'espace (ALICE) for the town of Saint-Prex on Lake Geneva, shown taking flight and opening the concert venue to the sky (courtesy Dieter Dietz, ALICE, EPFL).

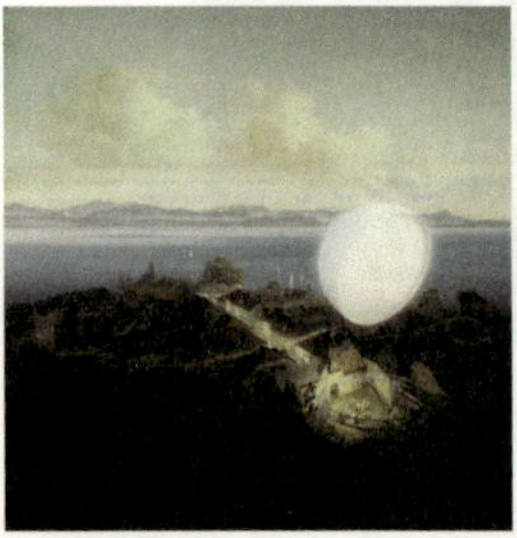

37 Exhibition catalogue *God & Co: François Dallegret Beyond the Bubble*, Alessandra Ponte, Laurent Stalder, and Thomas Weaver, eds. (London: AA Publications, 2011) and photographic proofsheet by François Dallegret entitled "Have a Ball! Have a Kick" from 1969.

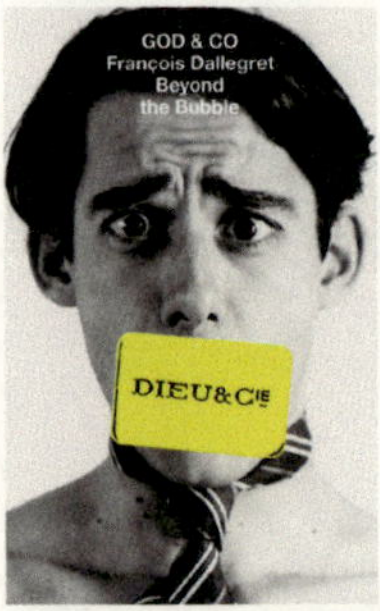

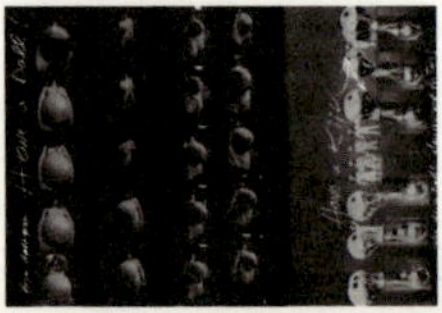

38 For a much earlier treatment of the issue of orbital debris, see *Life Magazine* from August 5, 1966, featuring the article by Albert Rosenfeld entitled "Space Trash."

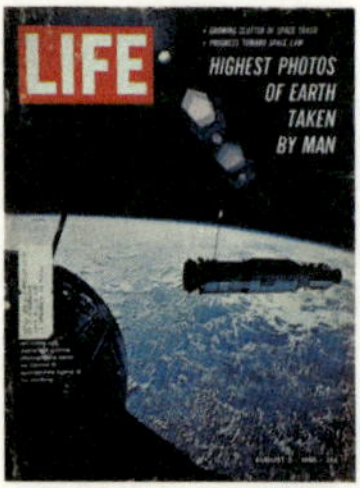

39 Distant future scenarios of fluctuations and polarity reversals of the Earth's magnetic field; Institute for Geophysics, Department of Earth Sciences, ETH Zurich, and Swiss National Supercomputing Centre, 2012 (courtesy Jon Rotvig, Andrew Jackson, and Jean Favre).

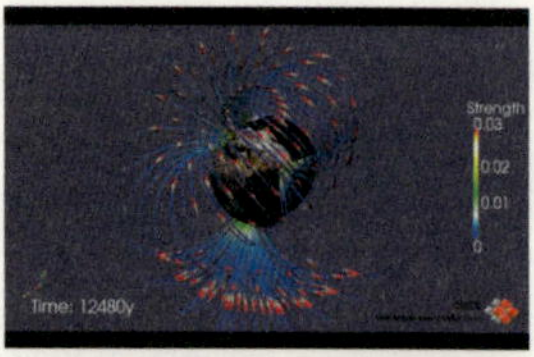

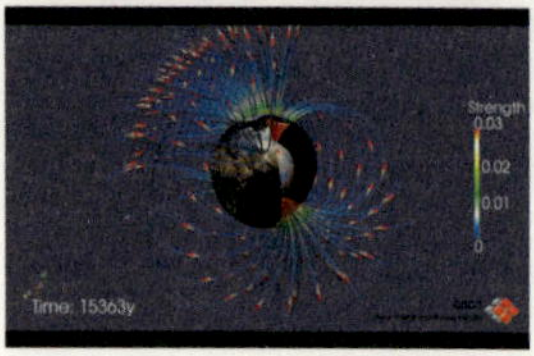

40 Installation *Terra Forming: Engineering the Sublime* by Adam Lowe and Jerry Brotton shown at the exhibition *Anthropocène Monument* in Toulouse in 2014-15; the installation was comprised of a 2x4 meter 3D relief model of the world in the process of being flooded to the height of Mt. Everest.

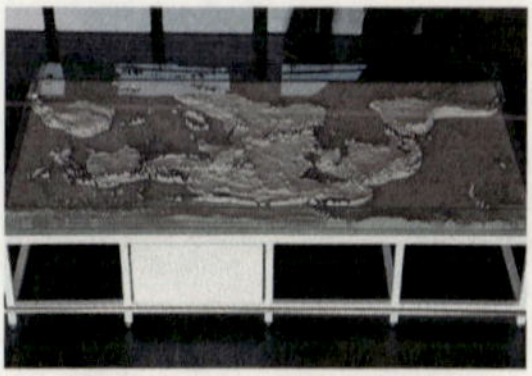

41 Sectional model of Earth at the *focus*Terra exhibition visualizing the geophysical fluid dynamics beneath the terrestrial surface, 2019 (courtesy Paul Tackley et al., Institute of Geophysics and *focus*Terra, ETH Zurich).

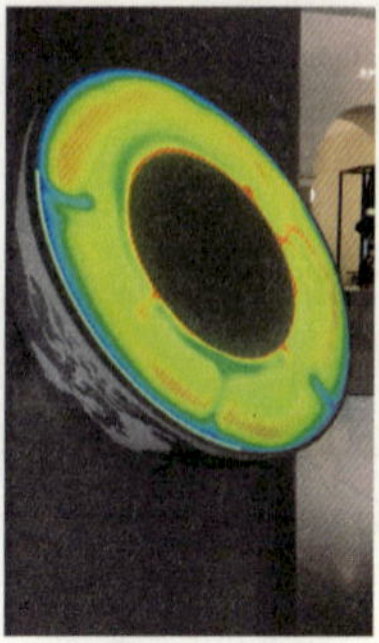

42 Cover of Parag Khanna's book *Connectography: Mapping the Future of Global Civilization*, 2016.

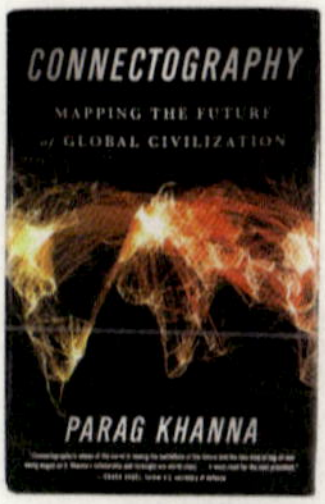

43 Web application demonstrating how maps are constructed, based on research conducted at the Institute of Cartography and Geoinformation of ETH Zurich, 2010-17 (courtesy Lorenz Hurni, Stefan Räber, et al.).

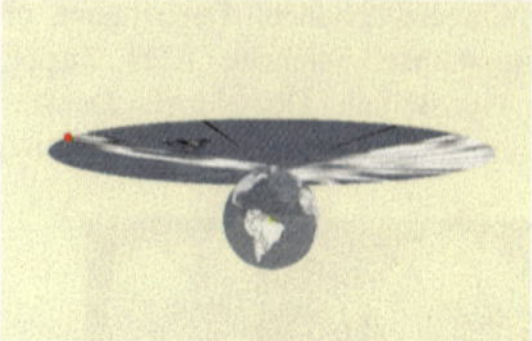

44 Book cover of *Geostories: Another Architecture for the Environment* by Design Earth (Rania Ghosn and El Hadi Jazairy), 2018.

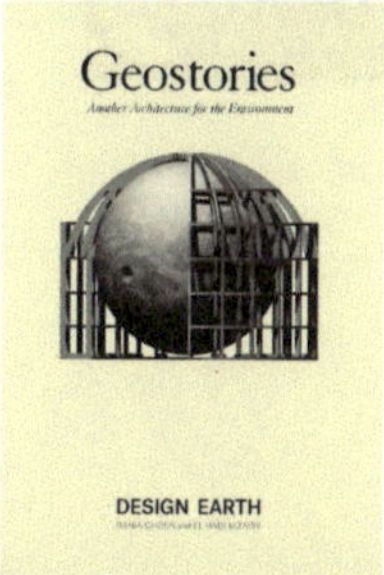

45 Sketch by architect Ciro Miguel completing Oscar Niemeyer's domed structure of the Ibirapuera Park exhibition building (1951-1954) with a submerged sphere; from the 2017 series *The Sunken Planets* (courtesy Ciro Miguel, Department of Architecture, ETH Zurich).

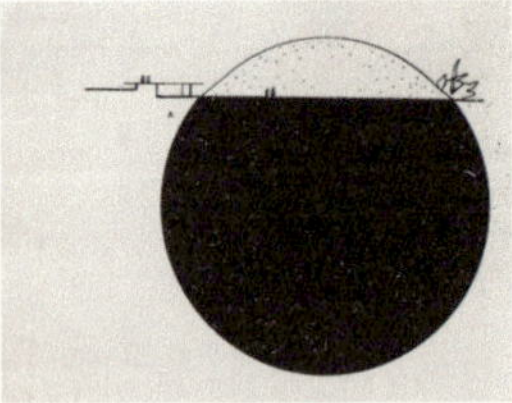

46 Video stills showing the spliting of the polar vortex into two vortices – an event known as Sudden Stratospheric Warming – that occurred in February 2018 (courtesy Institute of Atmospheric and Climate Science, Department of Environmental Sciences, ETH Zurich and Hochschule Düsseldorf; Daniela Domeisen and Alexander Wollert, 2018).

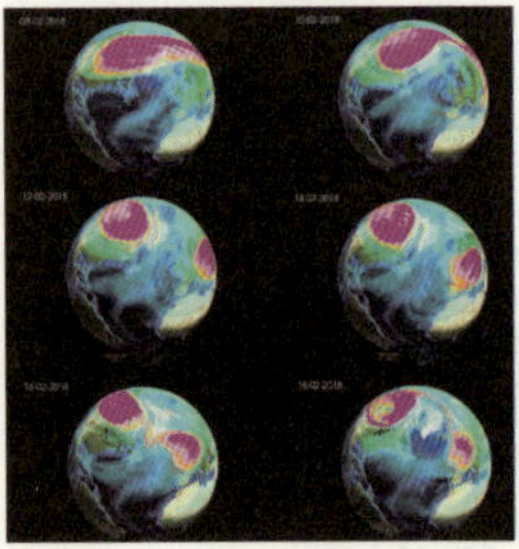

47 Image from a *Neue Zürcher Zeitung* article on the St. Galler Globus being photographed as the basis for producing a digital model of the historic globe; see Adi Kälin, "Der prächtige St. Galler Globus wird digitalisiert," *NZZ*, November 11, 2018 (photograph Nathalie Taiana, courtesy NZZ); and close-up image of the St. Galler Globus.

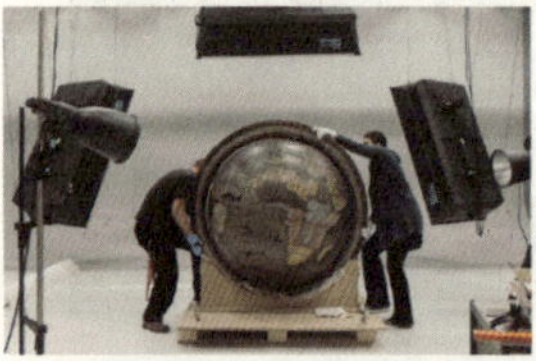

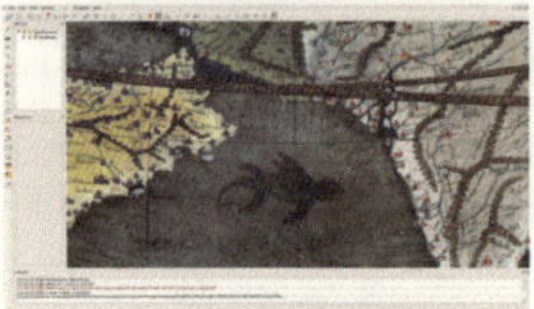

48 Digital model of the Earth geoid, showing the planet's gravitational field as an equipotential surface (courtesy Alain Geiger, Institute of Cartography and Geoinformation, Department of Civil, Environmental and Geomatics Engineering, ETH Zurich).

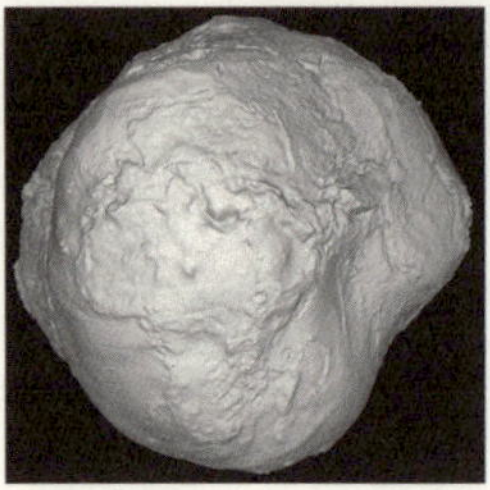

49 Conceptual diagram of processes driving the changes in the Southern Ocean carbon fluxes, comparing the situation in the 1990s with that in the 2000s (courtesy Nicolas Gruber et al., 2019, Environmental Physics Group, Institute of Biogeochemistry and Pollutant Dynamics, ETH Zurich).

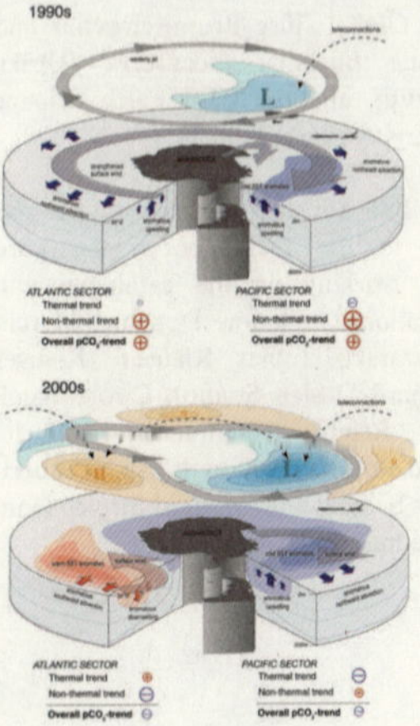

50 Cover of the *Brandt Report* by the Independent Commission on International Development Issues from 1980; the commission was chaired by former Chancellor of Germany Willy Brandt; and map showing the concentration of GDP in urbanized regions of the world as published in the *Global MetroMonitor 2014: An Uncertain Recovery*.

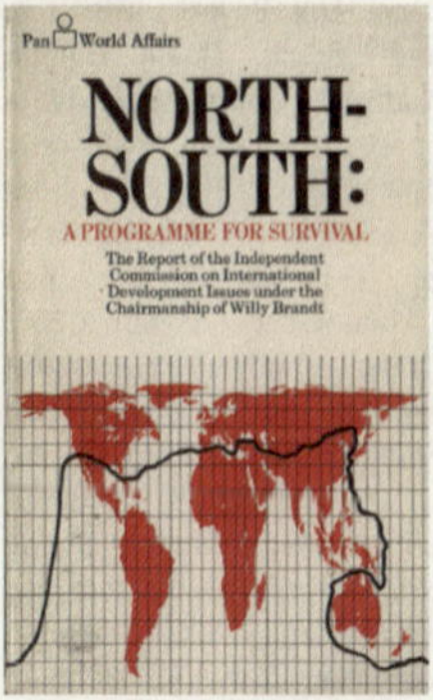

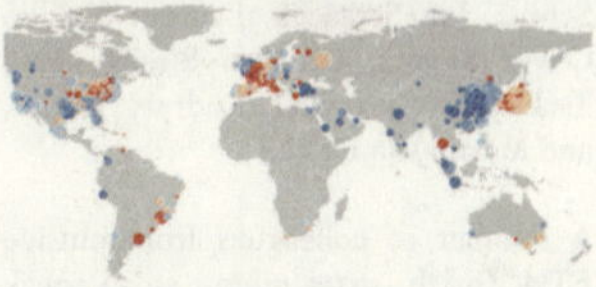

Acknowledgements

We would like to express our gratitude to the following colleagues from ETH Zurich for having answered our "Call for Globes," for their support in assembling the material for the booklet and exhibition *Terrestrial Tales*, and for having allowed us to share in their research endeavors, notwithstanding our limited knowledge of their particular fields of expertise:

Michael Afanasiev, Carolin Arndt Foppa, Kay Axhausen, Urs Beyerle, Christian Boehm, Peter Brack, David Bresch, Stephen Cairns, Giulia Celentano, Tarun Chadha, Kees Christiaanse, Daniela Domeisen, Jean Favre, Andreas Fichtner, Peter Fischer, Olivier Frey, Alain Geiger, Taras Gerya, Domenico Giardini, Nicolas Gruber, Armin Grün, Thomas Haag, Ludger Hovestadt, Lorenz Hurni, Andrew Jackson, Ulrike Kastrup, Frederick Kim, Reto Knutti, Lion Krischer, Alexander Lehnerer, Ciro Miguel, Andreas Müller, Stefan Räber, Johan Robertsson, Markus Rothacher, Jon Rotvig, Karin Sander, Konrad Schindler, Christian Schmid, Tobias Schmidt, Julian Schubert, Elena Schütz, Jan Sedlácek, Laurent Stalder, Oliver Stebler, Leonard Streich, Paul Tackley, Roman Walt, Andreas Wieser, and Martin van Driel.

A number of colleagues from outside ETH Zurich—*extra muros*, so to speak—enthusiastically sent us their work to be included in the collection of globes or helped us identify the necessary rights for publication. Our sincere thanks goes to them as well: Alexandra Arènes, Jon Berkeley, Neil Brenner, Jerry Brotton, Angelo Bucci, François Dallegret, Elizabeth Diller, Gabriel Dunne, Hans Frei, Rania Ghosn, Axelle Grégoire, Alice Jaeckel, El Hadi Jazairy, Scott Hessel, Nikos Katsikis, Parag Khanna, Ben Kirchner, Aaron Koblin, Rem Koolhaas, Gabriel Kozlowski, Bruno Latour, Adam Lowe, Jason W. Moore, Michael Najjar, Petra Noordkamp, Guido Nussbaum, Alessandra Ponte, Philippe Rekacewicz, Charles Renfro, Roi Salgueiro Barrio, Hashim Sarkis, Jost Schmid, Ricardo Scofidio, Madelon Vriesendorp, Thomas Weaver, Alexander Wollert, and Zoe Zenghelis.

We are thankful for the support received from gta Exhibitions (Fredi Fischli and Niels Olsen), ETH Zürich Hochschulkommunikation (Isabelle Herold), and ETH Global (Jürg Brunnschweiler and Simone Bucher), *focus*Terra (Ulrike Kastrup), and the ETH Earth Science Library (Andreas Müller).

Last but not least, a big thank you goes to the members of the core team working on the catalogue and exhibition: Guillaume Dopchie, Marcin Ganczarski, Tobias Klauser (Klauser Lienhard Design Studio), Ciro Miguel, Ellen Reinhard, Jenny Rodenhouse, David Roth, as well as Julian Schubert, Elena Schütz, and Leonard Streich from Something Fantastic.

Colophon

Authors: Marc Angélil and
Cary Siress, with Marcin Ganczarski,
Tobias Klauser, Ciro Miguel,
Ellen Reinhard, Jenny Rodenhouse,
David Roth, Julian Schubert,
Elena Schütz, Leonard Streich

Design: Something Fantastic Art
Department (Berlin)

Typeface: Life
Paper: Lux Cream 60 g
Printing: Druckhaus Nomos

Bibliographic information published by
Die Deutsche Bibliothek. Die Deutsche
Bibliothek lists this publication in
the Deutsche Nationalbibliographie;
detailed bibliographic data is available
online at http://dnb.de.

ETH
Eidgenössische Technische Hochschule Zürich
Swiss Federal Institute of Technology Zurich

Ruby Press
Schönholzer Str. 13–14
10115 Berlin
Germany

http://ruby-press.com

Printed in Germany
ISBN 978-3-944074-30-6